T0192281

Lecture Notes in Computer Science 14098

The series Lecture Notes in Computer Science (LNCS), including its subseries Lecture Notes in Artificial Intelligence (LNAI) and Lecture Notes in Bioinformatics (LNBI), has established itself as a medium for the publication of new developments in computer science and information technology research, teaching, and education.

LNCS enjoys close cooperation with the computer science R & D community, the series counts many renowned academics among its volume editors and paper authors, and collaborates with prestigious societies. Its mission is to serve this international community by providing an invaluable service, mainly focused on the publication of conference and workshop proceedings and postproceedings. LNCS commenced publication in 1973.

Victor Malyshkin
Editor

Parallel Computing Technologies

17th International Conference, PaCT 2023
Astana, Kazakhstan, August 21–25, 2023
Proceedings

 Springer

Editor
Victor Malyshkin (ID)
Institute of Computational Mathematics
and Mathematical Geophysics SB RAS
Novosibirsk, Russia

ISSN 0302-9743 ISSN 1611-3349 (electronic)
Lecture Notes in Computer Science
ISBN 978-3-031-41672-9 ISBN 978-3-031-41673-6 (eBook)
https://doi.org/10.1007/978-3-031-41673-6

Preface

The 17th International Conference on Parallel Computing Technologies (PaCT 2023) was a four-day event held in Astana, Kazakhstan. It was organized by the Institute of Computational Mathematics and Mathematical Geophysics of the Russian Academy of Sciences (Novosibirsk) in cooperation with Astana IT University (Astana, Kazakhstan), Novosibirsk State University, and Novosibirsk State Technical University.

The PaCT conference series started in Novosibirsk (Akademgorodok) in 1991 and has been held in various Russian cities mainly every odd year since then. The 15th Conference, PaCT 2019, took place in Almaty, Kazakhstan, and it was an honor to collaborate with our partners in Kazakhstan this year again. Since 1995, all the PaCT proceedings have been published by Springer in the LNCS series. The aim of the PaCT 2023 conference was to provide a forum for an exchange of views among the international community of researchers in the field of the development of parallel computing technologies. The PaCT 2023 Program Committee selected papers that contributed new knowledge in methods and tools for parallel solution of large-scale numerical simulation and data processing problems. The papers selected for PaCT 2023

- present results in development of automatic programming tools,
- study parallel implementation of basic data structures and algorithms,
- propose high-level HPC/HTC services and frameworks,
- investigate the problems of HPC systems management, such as job scheduling and monitoring.

Number of submitted papers: 23. They were subjected to a single-blind reviewing process. The average number of reviews per submitted paper was 2.4. The Program Committee selected 15 full papers for presentation at PaCT 2023.

Many thanks to our sponsors: the Ministry of Science and Higher Education of the Russian Federation, Russian Academy of Sciences, Ministry of Science and Higher Education of the Republic of Kazakhstan, and RSC Technologies.

August 2023   Victor Malyshkin

Organization

Organizing Committee

Conference Chair

V. E. Malyshkin   ICMMG SB RAS, NSU, NSTU, Novosibirsk, Russia

Conference Secretary

M. A. Gorodnichev   ICMMG SB RAS, NSU, NSTU, Russia

Organizing Committee Members

S. B. Arykov	ICMMG SB RAS, NSTU, Russia
M. A. Gorodnichev	ICMMG SB RAS, NSU, NSTU, Russia
S. E. Kireev	ICMMG SB RAS, NSU, Russia
A. E. Kireeva	ICMMG SB RAS, Russia
D. V. Lebedev	Astana IT University, Kazakhstan
Yu. G. Medvedev	ICMMG SB RAS, Russia
V. A. Perepelkin	ICMMG SB RAS, NSU, Russia
V. S. Timofeev	NSTU, Russia
G. A. Schukin	ICMMG SB RAS, NSTU, Russia
N. K. Zhakiyev	Astana IT University, Kazakhstan

Program Committee

Victor Malyshkin (Chairman)	Institute of Computational Mathematics and Mathematical Geophysics, Russian Academy of Sciences, Novosibirsk State University, Novosibirsk State Technical University, Russia
Sergey Abramov	Program Systems Institute, Russian Academy of Sciences, Russia
Farhad Arbab	Leiden University, The Netherlands
Jan Baetens	Ghent University, Belgium
Stefania Bandini	University of Milano-Bicocca, Italy
Thomas Casavant	University of Iowa, USA

Jou-Ming Chang	National Taipei University of Business, Taiwan
Hugues Fauconnier	University of Paris, The Research Institute on the Foundations of Computer Science (IRIF), France
Dan Feng	Huazhong University of Science and Technology, People's Republic of China
Juan Manuel Cebrián González	University of Murcia, Spain
Yuri G. Karpov	Peter the Great St. Petersburg State Polytechnic University, Russia
Alexey Lastovetsky	University College Dublin, Ireland
Jie Li	University of Tsukuba, Japan
Giancarlo Mauri	University of Milano-Bicocca, Italy
Igor Menshov	Keldysh Institute for Applied Mathematics, Russian Academy of Sciences, Russia
Dana Petcu	West University of Timisoara, Romania
Viktor Prasanna	University of Southern California, USA
Waleed W. Smari	Ball Aerospace & Technologies Corp., USA
Victor Toporkov	National Research University "Moscow Power Engineering Institute", Russia
Roman Wyrzykowski	Czestochowa University of Technology, Poland

Additional Reviewers

Sergey Kireev
Anastasia Kireeva
Vladislav Perepelkin
Georgy Schukin
Andrey Vlasenko

Sponsoring Institutions

Ministry of Education and Science of the Russian Federation
Russian Academy of Sciences
RSC Group

Contents

Algorithms

Distributed Systems Management

Automatic Programming and Program Tuning

Automation of Programming for Promising High-Performance Computing Systems

Vladimir Bakhtin⑩, Dmitry Zakharov⑩, Nikita Kataev(✉)⑩,
Alexander Kolganov⑩, and Mikhail Yakobovskiy⑩

Keldysh Institute of Applied Mathematics RAS, Moscow, Russia
`dvm@keldysh.ru`, `kaniandr@gmail.com`
`http://dvm-system.org`

Abstract. Automation of parallel programming may focus on various tasks the programmer is burdened while developing a parallel program. Related tools assist the program profiling and aid the programmer with transforming the program to a form suitable for the efficient parallelization. Finally, these tools express an implicit program parallelism using a chosen programming model and optimize the parallel program for target architectures. However, the choice of the target Application Programming Interfaces (API) is of great importance in the development of interactive parallelization tools. On the one hand, the perfect choice of API should ensure the programming of the variety of modern and promising architectures. On the other hand, API must simplify the development of assistant tools and allow the programmer to explore the decisions made by the automated parallelization system. System FOR Automated Parallelization (SAPFOR) is an umbrella of assistant tools designed to automate parallel programming. It accomplishes various tasks and allows the user to take an advantage of the interactive semi-automatic parallelization. SAPFOR expresses parallelism using the DVMH directive-based programming model, which aims at developing efficient parallel programs for heterogeneous and hybrid computing clusters. The paper presents an empirical study that examines the capability of SAPFOR to assist parallel programming on the example of development of a parallel program for numerical simulation of hydrodynamic instabilities.

Keywords: Automation of parallelization · Heterogeneous computational clusters · GPUs · Directive-based programming models · DVMH · SAPFOR

This work was supported by Moscow Center of Fundamental and Applied Mathematics, Agreement with the Ministry of Science and Higher Education of the Russian Federation, No. 075-15-2019-1623.

1   Introduction

Parallel program development is typically done with lower level APIs such as MPI, POSIX Threads, CUDA and OpenCL. Heterogeneous systems, which became a mainstream in past years, make the programmer use a variety of different APIs in a single parallel program. MPI allows the programmer to take advantage of distributed memory systems, POSIX Threads is suitable to utilize multicore compute nodes with shared memory, and CUDA and OpenCL provide the programmer with control over the GPUs. However, the application of lower level parallel programming models raises concerns from the complexity, correctness, and portability and maintainability perspectives. Portability is affected because programming of different architectures can be only accomplished using different API. For example, NVIDIA GPUs require CUDA while AMD GPUs rely on OpenCL to express parallelism. To address all mentioned concerns parallel programming approaches must raise the level of abstraction above that of lower level APIs.

A possible approach is to extend general-purpose languages with parallelism specifications expressed with higher level directives. This approach facilitates and unifies parallel programming for different architectures because the compiler is responsible for the implicit delivery of parallelism across different architectures using specific API. Directive-based programming models preserve the sequential code. Hence the ability for the normal compiler to neglect this kind of parallelism specifications greatly simplifies the introduction of new models of parallel programming.

An example of a model that supports multi-platform shared-memory parallel programming is OpenMP. The recent versions of the OpenMP standard propose an extension of the model for accelerator programming. Using this extension the programmer is capable to distribute parallelism across the whole cores (CPU and GPU) available in a single cluster node. A similar approach was implemented by Cray, NVIDIA and AMD in the OpenACC standard . It provides a set of compiler directives designed for developing parallel applications on both CPUs and GPUs. The arising of these higher level APIs significantly simplifies parallel programming. Now, instead of three programming models, only two are necessary to achieve a massive performance, for example MPI with OpenMP or MPI with OpenACC.

The DVMH [1] model, proposed at Keldysh Institute of Applied Mathematics, goes even further. In addition to API that provides the developer with a set of directives to distribute computations, the model also introduces specifications to facilitate interaction between cluster nodes. DVMH supports partitioning data between processors, manages accesses to remote data and implements global operations among data located on different nodes. Therefore, using DVMH the developer can completely abandon MPI like APIs. Parallel applications written in DVMH languages remain unchanged if they are transferred from one HPC system to another. The performance problem is addressed by the DVMH compilers and run-time system which implement various optimizations aimed at better utilization of target architecture.

While directive-based models reduce the programmer effort and increase software maintainability and portability, they are still explicitly parallel. Therefore, to purse high performance the programmer is burdened with the tasks of identifying code regions which mainly affect the application performance and revealing loop-carried data dependencies along with spurious dependencies, which do not prevent loop parallelization but require explicit specifications in a source code. Furthermore, in case of distributed memory systems the programmer chooses data to be partitioned between compute nodes and corresponding distribution rules, and finally specifies communication points in a source code to exchange data between different compute nodes. Thus, although explicit higher level APIs decrease the overall parallel programming complexity, they still complicate code writing and debugging. Introduction of interactive tools that perform program analysis and transformation is the next step towards the simplification of parallel programming.

In this paper we explore the capability of System FOR Automated Parallelization (SAPFOR) [1] to assist parallel programming on the example of development of a parallel program for numerical simulation of hydrodynamic instabilities. On the one hand, SAPFOR uses an implicit parallel programming methodology [2], which assumes that the auto-parallelizing compiler, it contains, turns the well-formed sequential program into a parallel one. On the other hand, SAPFOR extends an implicit parallel programming methodology and it provides the user with the interactive tools to examine the program at each step of parallelization and to obtain a well-formed version of a sequential program from the original one. This paper summarizes all previous works which form SAPFOR and presents an empirical study of its capabilities.

The rest of the paper is organized as follows. Section 2 briefly summarizes architecture and main capabilities of SAPFOR, it also discusses the use of the SAPFOR interactive subsystem in the process of parallelization. Section 3 in details describes parallelization of the program for numerical simulation of hydrodynamic instabilities. We highlight the main steps of parallelization and outline the contribution of SAPFOR at any of them. Section 4 presents the performance of the obtained parallel program. We examine its execution on a heterogeneous computational cluster and show how the choice of a launch configuration (number of CPU cores and GPUs) affects the program performance. Section 5 discusses the related work and, finally, Sect. 6 concludes this paper.

2   Architecture of SAPFOR

SAPFOR focuses on three main tasks:

- Exploration of sequential programs (program analysis and profiling).
- Automatic parallelization (according to the high-level programming model) of a well-formed program for which a programmer maximizes algorithm-level parallelism and asserts high-level properties (implicit parallel programming methodology).
- Semi-automatic program transformation to obtain a well-formed sequential version of the original program.

We can distinguish three major parts of SAPFOR that focus on solving these tasks: the core subsystem, that manages program analysis and transformation, the dynamic analysis library and the interactive subsystem. These parts are independent from each other and communicate using specific APIs, that allow us to use different implementations of interactive subsystem and dynamic analyzers if necessary.

The integration between core and interactive subsystem [3] relies on the client-server model implemented using message passing interface with messages encoded in JSON. The interactive subsystem can be implemented as a standalone Integrated Development Environment (IDE) or as an extension for widely used tools. In this study we use the implementation of the interactive subsystem as an extension for the Visual Studio Code editor.

SAPFOR implements instrumentation-based dynamic analysis [6], hence the dynamic analysis library is an external library that implements a specific interface functions. Dynamic analysis results are encoded in JSON.

The core subsystem is organized as a pass framework. Passes perform analysis and transformation of the program. To implement new capabilities the number of existing passes can be extended. Another major part of the core subsystem is a multi-level representation of the program. It includes three levels. At the low level LLVM IR [4] is used to perform program analysis and property sensitive transformations that bring the low level program representation to the most suitable form for program analysis. At the top level Clang AST and Flang AST are used to perform source-to-source transformation of the original program and to navigate the programmer through the analysis results in a user-friendly way. At the middle level extended metadata-based representation is used to establish correspondence between lower level IR and higher level AST.

Data dependencies between statements determine the information structure of a sequential program [5]. Hence from the SAPFOR perspective we need to focus on memory accesses to exploit implicit parallelism. Analysis of memory accesses addresses recognition of data dependencies that prevent program parallelization, as well as spurious data dependencies (reduction and induction variables, privatization techniques), alias analysis and points-to analysis, recognition of pointers captured in function calls, data flow analysis, etc. The program representation influences the quality of analysis results. Hence we interested in application analysis techniques after specific transform sequences, which may include but not limited to construction of Static Single Assignment (SSA) form, dead code elimination, function inlining, combining redundant instructions, natural loop canonicalization and etc. Moreover distinct pass sequences are better suited to different properties to be analyzed.

SAPFOR performs mentioned transformations on top of the low level program representation and does not affect the source code. To bring the gap between LLVM IR and AST-based representations we proposed a novel data structure, called source-level alias tree [7].

The source-level alias tree allows SAPFOR to establish correspondence between IR-level memory locations accessed in a transformed program and AST-level memory locations accessed in the original program. Thus, analysis in SAP-

FOR is organized as a number of concurrent pass sequences (analysis servers) [8]. Each sequence contains analysis and transform passes that are better suited to investigate required program property. We combine multiple analysis results using a source-level alias tree and, finally, obtain the overall analysis of the program.

A separate passes were implemented to perform source-to-source transformations, parallelization for shared memory systems (multicore CPUs and GPUs) [8] and parallelization for distributed memory systems [9].

If SAPFOR relies on DVMH model to express loop-level parallelism, it automatically generate the following specifications to get a resulting parallel program: (1) specifications for data distribution, (2) specifications of the loops which can be executed in parallel, as well as specifications of private and reduction variables, and a pattern of array accesses, (3) specification of the compute regions which can be executed on the accelerators, each region may enclose one or more parallel loops, (4) high-level specifications of data transfer between a memory of CPU and a memory of accelerator (actualization directives).

Here is a brief description of the major stages of the parallelization strategy:

1. Clang and Flang are used as frontends to parse the sequential program into an intermediate representation. The user uses the interactive subsystem to form a project to be analyzed and to specify compiler options required to parse source code. SAPFOR has the ability to use a compilation database which specifies how translation units are compiled in the project. Some build tools like CMake support generation of compilation databases.
2. The programmer uses a DVMH profiling tool, which is a part of the DVM system, or other conventional tools like gprof and llvm-cov to emphasize source code regions that should be considered in parallelization. SAPFOR uses static and dynamic techniques to assess the possible parallel program performance and to determine the best parallelization strategy. The programmer is also able to use SAPFOR instrumentation-based analysis techniques to maximize the overall accuracy of the program analysis.
3. The interactive subsystem allows the programmer to observe analysis results and to make assertions about program properties that SAPFOR has been failed to analyze.
4. If the original computations are not parallelizable as given the programmer chooses essential transformations and regions of a source code to be transformed. SAPFOR checks precondition to approve that desirable transformations do not affect the original program behavior. If automatic transformation is successful, the parallelization continues from steps (2) or (3). Otherwise, the program should be transformed in a manual way.
5. Automatic parallelizing compiler, which is a part of SAPFOR, explicitly expresses parallelism in a source code using preferred parallel programming API (DVMH or OpenMP are supported at the moment). The problem of distributed memory parallelization requires a solution to three main subproblems: data and computation distribution and communication optimization [9]. To reduce parallelization overheads SAPFOR explores all memory accesses in the entire program while it solves the data partitioning sub-

problem. To reduce the complexity of the problem, which is generally NP-hard, the programmer can specify regions of the source code to resolve separately. Finally, the programmer needs to join the obtained solutions in a manual way to find resulting solution for the entire program.

3   Development of a Parallel Program

In this section we follow the major stages in the parallelization of a software package for numerical simulation of hydrodynamic instabilities. The paper [10] discusses algorithms used for calculation of hydrodynamic. Various modifications of this package are developed and used for scientific calculations at Keldysh Institute of Applied Mathematics RAS [11–13].

3.1   Original Program Profiling

We conducted a performance evaluation to extract regions of code to be considered in parallelization and to estimate the impact of optimizations made by state-of-the-art compilers.

The results were gathered on the system equipped with Intel Xeon CPU E5-1660 v2, 3.70 GHz. The application was compiled with Intel C/C++ Compiler version 19.0.2.187 with option -O3. We used representative test data with the reduced grid size (3000×1257) and the reduced number of iterations (10) to collect results within a reasonable time. The program execution time was 266 s (4.5 min). Computations in one function, namely *compute_it*, takes 95.6% of the program execution time. Further analysis showed that this function consists of two independent parts of computations (functions *compute_heat* and *compute_hydro*) that takes 70% and 30% of the function execution time correspondingly. I/O of intermediate results takes the rest of the program execution time.

This performance study showed that the main computations are organized in the form of loop. Thus we decided to use capabilities of the DVMH model to express loop-level parallelism.

3.2   Original Program Analysis

The original program totals about 10000 lines of code in the C language. It contains 187 functions, including 137 user-defined functions, functions from the C standard library and the GNU Scientific Library [14]. User-defined functions declare 1617 variables and comprise 244 loops. However, SAPFOR found direct calls of 101 functions only. SAPFOR also found 11 indirect function calls, hence there may be more functions involved in the computations. The total amount of calls from user-defined functions is 612. Absence of LLVM IR representation for precompiled library functions prevents SAPFOR from ascertaining the total number of function calls and corresponding callees.

The time-consuming part of the code involves 82 directly called functions (414 call statements from *compute_it* and its descendant functions in the program call graph): 29 functions are called from *compute_hydro* (93 call statements, 2 indirect calls, 22 user-defined functions), 18 functions are called from *compute_head* (94 call statements, 7 indirect calls, 10 user-defined functions).

Initially, SAPFOR recognized only 66 loops without significant parallelization issues (loop-carried data dependencies, I/O statements, multiple loop exits, statements that lead to the unexpected program termination). Unfortunately, the most important loops were not fully analyzed due to the presence of indirect function calls and inaccuracy of the alias analysis. SAPFOR was not managed to disambiguate formal parameters which points to the beginning of different arrays. As a result, it made an assumption of indirect accesses to these arrays.

To resolve these issue we transform the original program in a way discussed in the next section and perform dynamic analysis to investigate memory accessed from the indirect function calls.

3.3   Original Program Transformation

In this section we overview source-to-source transformations we made to improve the accuracy of points-to analysis.

In the program different array parameters of a function refers to different physical quantities. Each element of the array represents a compute state in a corresponding grid point. Thus, different array parameters point to different memory locations and do not alias. We use the *restrict* qualifier, which is natural in C language, and helps SAPFOR to properly compute alias relations. We manually changed prototypes of several functions and added 15 *restrict* qualifiers.

However, when using arrays of pointers, the *restrict* qualifier is not enough to disambiguate memory accesses to formal parameters. The C language turns accesses to these arrays into sequences of two dereference statements. In this case, the *restrict* qualifier does not affect the second statement that refers a pointer stored inside the array. We implemented a demand-driven source-to-source transformation in SAPFOR aimed at splitting small arrays of pointers into independent variables. Each element of the original array-parameter results in an independent parameter of a pointer type, hereafter we can apply the *restrict* qualifier.

There are two steps in the transformation:

1. The programmer specifies parameters to replace. SAPFOR creates a copy of an original function, with a modified set of formal parameters. It also replaces accesses to original parameters in the body of new function.
2. The programmer specifies function calls to replace with a copy of the callee which has been created in the first step.

Listing 1.1 shows the first step of the transformation. The SAPFOR directive *replace* specifies a parameter to replace in the function *foo*. Its copy *foo_spf0* with a new prototype is created and all accesses to *ss[0]* and *ss[1]* are replaced with new parameters *ss_ 0* and *ss_ 1* correspondingly. The *nostrict* clause has a meaning similar to the *restrict* qualifier. In the example it ensures SAPFOR that all accesses to *ss* in the *foo* function occur through *ss*. Without this clause the transformation was still possible, however SAPFOR would pass a pointer to each element of *ss* to the *foo_ spf0* function. Therefore like in the original program, two dereference statements would form accesses to values stored in the array.

Listing 1.1. Automated replacement of an array-parameter in a function

```
void foo(state_t **ss) {
#pragma spf transform replace(ss) nostrict
   /* accesses to ss[0] and ss[1] */
}

/* Replacement for void foo(state_t **ss) */
void foo_spf0(state_t *ss_0, state_t *ss_1) {
#pragma spf transform metadata \
                      replace(foo, { .0 = ss_0, .1 = ss_1})
   /* accesses to ss_0 and ss_1 */
}
```

Listing 1.2 shows the second step of the transformation. SAPFOR replaces a call to *foo* with a call to *foo_ spf0*.

Listing 1.2. Automated replacement of the calling function

```
void bar(state_t **ss) {
   #pragma spf transform replace with(foo_spf0)
   foo(ss);
}

void bar() {
   foo_spf0(ss[0], ss[1]);
}
```

3.4   Dynamic Analysis

At the next step we checked data dependencies in the program at runtime. Then we complemented static analysis results with the gathered data. Blended analysis of the transformed program allows SAPFOR to find 95 loops without significant parallelization issues including important loops that spend the most part of the program's execution time. Some loops with indirect function calls were also chosen for parallelization. These loops can be parallelized for multicore CPU, however parallelization for distributed memory and GPU requires additional parallelism specifications that cannot be applied to indirect function calls.

To reduce dynamic analysis time we instrumented important functions only, including functions which are called indirectly. SAPFOR provides a programmer with an option to select the starting points for instrumentation. It uses a program call graph to only instrument selected functions and connected descendant functions. We also explicitly select functions which are called indirectly. Pointers to these functions are initialized at the program beginning before any computations are started. Input data determines particular functions to be called at runtime. Thus, it was enough to select all the functions, pointers to which can be used.

We explicitly selected 14 functions, including 10 indirectly called functions. As a result, SAPFOR instrumented 75 functions instead of 139, which are instrumented if it analyzes the entire program. We used reduced grid size (150×63) and profiled 1 iteration only. Our measurements were taken on a desktop workstation which consists of Intel Core i7-10510U, 1.8 GHz CPU. The total analysis time was 409 sec. (6 min. 49 sec.), the memory consumption during the analysis was 336 MB. The program execution time without using dynamic analysis is 0.2 sec. Thus, dynamic analysis slows down the program up to 2045 times. For comparison, the time of static analysis of the program is 102 sec.

3.5 Parallelization for Shared Memory

SAPFOR was capable to recognize important loops without significant parallelization issues after the step mentioned in the previous sections. However, parallelization for GPUs required additional program transformations.

Firstly, it is not allowed to take address of a function in a host code and to use this address in device code. We manually replaced 10 indirect function calls with direct calls to enable parallelization of important loops. Secondly, calls to the GSL library within the loop body prevent parallelization for GPUs, because there is no GPU conformance version of the library available. In total 184 GSL library calls were restructured and 20 GSL vectors were replaced with natural C arrays. Finally, CDVMH language, which is a target API SAPFOR uses, imposes some restrictions on functions that is called in regions of code which are scheduled for GPU execution [15]. These functions cannot produce side effects and accesses remote data which are located on another processor. SAPFOR performed inline substitution of these functions in the source code.

The application of the mentioned restructuring techniques made it possible to construct DVMH version of the original program suitable for execution on multicore CPUs and GPUs.

3.6 Parallelization for Distributed Memory

Unlike incremental parallelization applicable for shared memory, distributed memory requires global decision making. Three main sub-problems need to be addressed: data and computation distribution and communication optimization. Unfortunately, linearized multidimensional arrays and program modularity drastically complicate data distribution.

It is natural to use a multidimensional grid to represent a multidimensional space and a multidimensional array to represent a state of computations at any point in the grid. The DVMH model relies on a multidimensional view of a distributed memory system and it provides the programmer with explicit specifications to map data and computations to a node in a multidimensional grid of virtual compute nodes. If the main source of parallelism is nested loops, the *parallel* directive specifies compute nodes to execute iterations on. An entire iteration of a loop can be executed by a single node only. Hence to improve data locality and to reduce communication overhead a compute node need to allocate all data an iteration accesses. Thus, there is a relation between elements of different arrays. In the source code the *align* directive expresses this relation. It establishes the arrangement of array elements relative to each other on the multidimensional view of the compute system. To specify alignment affine expressions in a form $a*i+b$ is used for each dimension of an array.

Linearized view of data corresponds to the array representation in memory, but hides the multidimensional structure of data and hinders data partitioning. SAPFOR implements techniques to recover the form of multidimensional arrays in the C99 language, which presented in lower level LLVM representation in a linearized form [16]. While SAPFOR can recover the original form of multidimensional arrays, a source-to-source transformation that turns linearized arrays into multidimensional ones is not implemented yet. It can be done only by looking at all array accesses in an entire program and it requires complex interprocedural analysis.

The original program is structured into modules. This modularity raises the compiler issue of interprocedural analysis because it is not known at compile time which of the functions will be called during a specific execution. Different modules require different data partitioning and some of the modules cannot be parallelized because of loop-carried data dependencies or unstructured control flow.

Thus, for the entire program, SAPFOR was not able to construct the data distribution in an automatic way. However, SAPFOR implements incremental parallelization techniques for distributed memory [1]. We applied SAPFOR to independently solve data distribution problem for the explicitly selected well-formed regions of source code. The found solutions helped us to manually select data distribution for the entire program.

3.7 Parallel Program Optimization

The main drawback the parallel program has at this step was the presence of imperfect loop nests in the *compute_ heat* function that spend for than 90% of the program's execution time. As a result, it was not possible to collapse iteration spaces of two nests of loops to increase loop-level parallelism. We applied several code reconstruction techniques to obtain perfectly nested loops.

Firstly, SAPFOR performs automatic function inlining in loops that form a loop nest. Secondly, we expand a number of dimensions of 10 arrays which were used temporarily within an iteration of an outer loop. The transformation makes each iteration in the nest access a separate element of an array and breaks data dependencies in the loop. We did this reconstruction in a semi-automatic way. We applied these two transformations to 2 loops. Each of these loops had had 2 directly nested inner loops, a call of a function that contains 3 loops and standalone statements. As a result, we obtained 14 loop nests with each nest containing 2 perfectly nested loops. The transformation enables the partitioning of computations in the program between nodes of a two-dimensional grid of cluster nodes.

Finally, all discussed steps follow us to a parallel DVMH program that is suitable for execution on heterogeneous computation cluster with accelerators. The resulting program totals 21450 lines of code of which DVMH directives occupy about 500 lines. The following specifications have been inserted:

- 107 directives to specify parallel loop nests (*parallel*),
- 20 directives to specify execution on GPU (*region, get_ actual, actual*),
- 160 data distribution directives (*distribute, align, realign, redistribute*),
- 66 directives to specify functions that inherit data distribution from the caller function (*inherit*),
- 5 directives to specify accesses to remote data (*remote_ access*).

4 Parallel Program Performance Evaluation

We have measured the performance of the developed parallel program on the K60 supercomputer of Shared Resource Center of KIAM RAS [17]. Figure 1 and Fig. 2 show the execution time we observed depending on number of CPUs and GPUs.

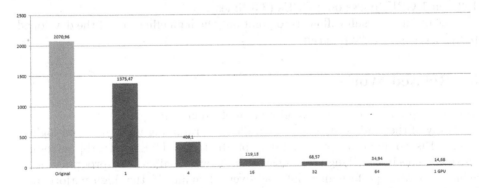

Fig. 1. The execution time (sec.) of 100 iterations on the grid 3000×1528

Source code reconstruction techniques discussed in previous sections reduce the program execution time on a single core. On the 3000×1528 grid the program speedup is of about 1.5. On 64 cores of two Intel Xeon Gold 6142v4 processors it runs up to 59 faster than the original program. A single NVIDIA Tesla V100 GPU is enough to achieve speedup of about 141. Figure 2 shows fragments of the output of the performance analysis tool which is a part of the DVM system. The figure compares the main characteristics that form program execution time on 12000×5028 grid if 1 and 4 GPS are used correspondingly.

```
1 GPU   (Tesla V100-PCIE-32GB)

                            #        Min       Max       Sum      Average   Productive      Lost

[Region IN] Copy CPU to GPU  3146      4B      1.350G    25.205G    8.204M     2.2477s         -
Loop execution               5600    0.0000    0.2657   158.7707    0.0284   158.7707s         -
Reduction                     300    0.0006    0.0013     0.1837    0.0006         -        0.1837s
Page lock host memory        6200    0.0000    0.0029     0.7681    0.0001         -        0.7681s

Productive time:   161.0184s
Lost time      :     0.9517s


4GPU #  (Tesla V100-PCIE-32GB)

                            #        Min       Max       Sum      Average   Productive      Lost

[Shadow] Copy GPU to CPU     9620   19.664K   140.672K  416.625M   44.347K        -        0.5162s
[Shadow] Copy CPU to GPU     7386   19.664K   281.344K  513.096M   71.136K        -        0.4169s
[Region IN] Copy CPU to GPU  3214      4B     346.047M    6.308G    2.010M     1.1567s         -
GET_ACTUAL                    110   115.349M  346.047M   16.897G   157.294M    2.7507s         -
Loop execution               4200    0.0000    0.0607    38.9186    0.0093    38.9186s         -
Reduction                     300    0.0002    0.0008     0.0565    0.0002         -        0.0565s
Page lock host memory        6294    0.0000    0.1056     4.6534    0.0007         -        4.6534s

Productive time:    42.8260s
Lost time      :     5.6431s
```

Fig. 2. Performance profiling on 1 and 4 GPUs (grid size 12000×5028)

On 4 GPUs the overhead increases (the *lost time* field) up to 5.64 s in comparison with 0.95 s on 1 GPU. Exchange of shadow edges between different GPUs incurs the overhead of about 0.92 s. However, despite the increasing overhead 4 GPUs improve the program performance and reduces the execution time from 162 s on 1 GPU to 48 s on 4 GPUs (3.3 times).

The obtained results allow us to point out the high efficiency of the developed parallel version of the program.

5 Related Works

Various approaches exist to simplify parallel programming.

Many of them advocate explicit parallel programming while relies on higher-level APIs to increase programming productivity. DSLs [18,19] introduce narrowly specialized language constructs, however allow the programmer to achieve greater performance into a given domain. If the high-performance libraries [20,21] are used, a general-purpose programming languages are enough to utilize parallel platforms. However, the programmer is limited by the capabilities a library implements and by parallel architectures a library supports. Directive-based programming models are general enough to express parallelism in different ways, but they stile require a developer to be very aware of parallel programming.

Another group of approaches proposes assistance tools which address different steps of the parallelization process. Some of them do not make decision of how to parallelize loops but perform data dependence profiling [22] or make recommendations to the user which code regions to parallelize first [23]. However, the most desirable tools are automatic parallelizing compilers.

Model-based parallelizing compilers [24–27] represent the entire transformation sequence as a single transformation and discover it in an automatic way using mathematical optimization methods. Usually these compilers implement incremental parallelization techniques and utilize shared memory systems only. Moreover, a complex mathematical model which is used to represent the program fragments drastically complicates user participation in the parallelization.

Tools that target distributed memory systems do not usually overcome all sub-problems: data and computation distribution, communication optimization. These compilers may derive computation distribution from predefined data distribution [28, 29] or introduce special data types to simplify decision making [30].

If the program parallelization require preliminary transformations of a source code or a compiler fails to analyze the program properties, the compiler generates code which is not optimal. Some tools provide the user with capabilities to participate in the parallelization process using an interactive subsystem [28] or asserting program properties in a source code [31].

6   Conclusion

Many sequential applications have been written since the appearance of computer systems. These programs form the basis for the development of parallel programs. However, it is apparent that automatic parallelization for heterogeneous and hybrid computing cluster cannot be applied to large-scale codes yet. In this paper we point out main steps in the parallelization process that current assistant tools should focus on to make semi-automatic parallelization to become beneficial to developers of large-scale applications. We summarize the experience of parallel program development we gain in Keldysh Institute of Applied Mathematics RAS to propose a methodology the developers can follow to overcome the main parallel programming issues: complexity, correctness, and portability and maintainability.

We proposes an automation system (SAPFOR) and parallel API, which explicitly expresses parallelism in a source code (DVMH), that complement each other to empower opportunity of semi-automatic parallel program development. SAPFOR implements various techniques to assist any step of the parallelization process: sequential program profiling, extraction of the original program properties essential for its parallelization, reconstruction of the source code, data and computation distribution as well as communication optimization.

DVMH programs can be executed without any changes on workstations and HPC systems equipped with multicore CPUs, GPUs, and Intel Xeon Phi coprocessors. The performance gains, which are achieved on different architectures, are caused by various optimizations implemented in the DVMH compiler and runtime system. At startup the programmer configures desirable resources (the number of cluster nodes, threads and accelerators, the number of processors per node as well as performance of different processing units) the parallel application should utilize. Thus the best configuration can be selected to improve the efficiency of computational resources utilization in HPC centers.

References

1. Bakhtin, V.A., Krukov, V.A.: DVM-approach to the automation of the development of parallel programs for clusters. In: Programming and Computer Software, vol. 45, no. 3, pp. 121–132 (2019) https://doi.org/10.1134/S0361768819030034
2. Hwu, W.-M., et al.: Implicitly parallel programming models for thousand-core microprocessors. In: Proceedings of the 44th annual Design Automation Conference (DAC '07), pp. 754–759. ACM, New York, NY, USA (2007). https://doi.org/10.1145/1278480.1278669
3. Kataev, N.: Interactive Parallelization of C Programs in SAPFOR. In: Scientific Services & Internet 2020. In: CEUR Workshop Proceedings, vol. 2784, pp. 139–148 (2020)
4. Lattner, C., Adve, V.: LLVM: A Compilation Framework for Lifelong Program Analysis & Transformation. In: Proceedings of the 2004 International Symposium on Code Generation and Optimization (CGO'04). Palo Alto, California (2004)
5. Voevodin, V.V.: Information structure of sequential programs. Russ. J of Num. An. Math Modell. **10**(3) 279–286 (1995)
6. Kataev, N., Smirnov, A., Zhukov A.: Dynamic data-dependence analysis in SAPFOR. In: CEUR Workshop Proceedings, vol. 2543, pp 199–208 (2020)
7. Kataev, Nikita: Application of the LLVM compiler infrastructure to the program analysis in SAPFOR. In: Voevodin, Vladimir, Sobolev, Sergey (eds.) RuSCDays 2018. CCIS, vol. 965, pp. 487–499. Springer, Cham (2019). https://doi.org/10.1007/978-3-030-05807-4_41
8. Kataev, N.: LLVM based parallelization of C programs for GPU. In: Voevodin, V., Sobolev, S. (eds.) RuSCDays 2020. CCIS, vol. 1331, pp. 436–448. Springer, Cham (2020). https://doi.org/10.1007/978-3-030-64616-5_38
9. Kolganov, A.S., Kataev, N.A.: Data distribution and parallel code generation for heterogeneous computational clusters. In: Proceedings of the Institute for System Programming of the RAS (Proceedings of ISP RAS), vol. 34, no. (4), pp. 89–100 (2022) https://doi.org/10.15514/ISPRAS-2022-34(4)-7
10. Tishkin, V.F., Nikishin, V.V., Popov, I.V., Favorski A.P.: Finite difference schemes of three-dimensional gas dynamics for the study of Richtmyer-Meshkov instability (in Russian), vol. 7, no. 5, pp. 15–25 (1995)
11. Ladonkina, M.E.: Numerical simulation of turbulent mixing using high performance systems. PHD Thesis, Institute for Mathematical Modelling RAS (2005)
12. Kuchugov, P.A.: Dynamics of turbulent mixing processes in laser targets. PHD Thesis, Keldysh Institute of Applied Mathematics RAS (2014)
13. Kuchugov, P.A.: Modeling of the implosion of thermonuclear target on heterogeneous computing systems (in Russian). In: Proceedings of international conference Parallel computational technologies (PCT'2017), pp. 399–409. Publishing of the South Ural State University, Chelyabinsk (2017)
14. GSL - GNU Scientific Library. https://www.gnu.org/software/gsl/ Last Accessed 6 May 2023

15. C-DVMH language, C-DVMH compiler, compilation, execution and debugging of DVMH programs. http://dvm-system.org/static_data/docs/CDVMH-reference-en.pdf Last Accessed 6 May 2023
16. Kataev, N., Vasilkin, V.: Reconstruction of multi-dimensional arrays in SAPFOR. In: CEUR Workshop Proceedings, vol. 2543, pp. 209–218 (2020)
17. Heterogeneous cluster K60. https://www.kiam.ru/MVS/resourses/k60.html. Last Accessed 6 May 2023
18. Beaugnon, U., Kravets, A., Sven van Haastregt, Baghdadi, R., Tweed, D., Absar, J., Lokhmotov, A.: Vobla: A vehicle for optimized basic linear algebra. In: Proceeidngs of the 2014 SIGPLAN/SIGBED Conference on Languages, Compilers and Tools for Embedded Systems, LCTES '14, pp. 115–124, New York, NY, USA (2014)
19. Zhang, Y., Yang, M., Baghdadi, R., Kamil, S., Shun, J., Amarasinghe, S.: Graphit: A high-performance graph dsl. In: Proceedings ACM Program. Lang., 2(OOPSLA), pp. 121:1–121:30 (2018)
20. An, P., et al.: STAPL: an adaptive, generic parallel C++ library. In: Dietz, Henry G.. (ed.) LCPC 2001. LNCS, vol. 2624, pp. 193–208. Springer, Heidelberg (2003). https://doi.org/10.1007/3-540-35767-X_13
21. Bell, N., Hoberock, J.: Thrust: A Productivity-oriented library for CUDA. In: GPU Computing Gems, Jade Edition, Edited by Wen-mei W. Hwu, pp. 359–371 (2012). https://doi.org/10.1016/B978-0-12-385963-1.00026-5
22. Kim, M., Kim, H., Luk, C.-K.: Prospector: a dynamic data-dependence profiler to help parallel programming. In: 2nd USENIX Workshop on Hot Topics in Parallelism (HotPar '10) (2010)
23. Garcia, S., Jeon, D., Louie, C., Taylor, M.B.: Kremlin: rethinking and rebooting gprof for the multicore age. In: ACM SIGPLAN Notices June (2011). https://doi.org/10.1145/1993316.1993553
24. Bondhugula, U., Hartono, A., Ramanujam, J., Sadayappan, P.: A practical automatic polyhedral parallelizer and locality optimizer. SIGPLAN Notices 43(6), 101–113 (2008)
25. Verdoolaege, S., Juega, J. C., Cohen, A., Gomez, J. I., Tenllado, C., Catthoor, F.: Polyhedral parallel code generation for CUDA. ACM Trans. Archit. Code Optim. 9(4), 1–23 (2013)
26. Grosser, T., Groesslinger, A., Lengauer. C.: Polly – performing polyhedral optimizations on a low-level intermediate representation. Parallel Process. Lett. 22(04), 1250010 (2012)
27. Grosser, T., Hoefler, T.: Polly-ACC Transparent compilation to heterogeneous hardware. In: ICS '16: Proceedings of the 2016 International Conference on Supercomputing June 2016, pp. 1–13 (2016). https://doi.org/10.1145/2925426.2926286
28. Zima, H., Bast, H., Gerndt, M.: SUPERB: a tool for semi-automatic MIMD/SIMD parallelization. Parallel Comput. 6, 1–18 (1998). https://doi.org/10.1016/0167-8191(88)90002-6
29. Amarasingh, S. P., Lam, M. S. Communication Optimization and Code Generation for Distributed Memory Machines. In: PLDI '93: Proceedings of the ACM SIGPLAN 1993 conference on Programming language design and implementation, pp. 126–138 (1993) https://doi.org/10.1145/155090.155102
30. Kruse, M.: Introducing Molly: distributed memory parallelization with LLVM. CoRR, vol. abs/1409.2088 (2014). https://doi.org/10.48550/arXiv.1409.2088
31. Vandierendonck H., Rul S., Koen De Bosschere. The Paralax infrastructure: automatic parallelization with a helping hand. In: Proceedings of 2010 19th International Conference on Parallel Architectures and Compilation Techniques (PACT), IEEE, pp. 389–400 (2010). https://doi.org/10.1145/1854273.1854322

Automatic Parallelization of Iterative Loops Nests on Distributed Memory Computing Systems

A. P. Bagliy$^{(\boxtimes)}$, E. A. Metelitsa , and B. Ya. Steinberg

Southern Federal University, Rostov-on-Don, Russia
{abagly,elmet,byshtyaynberg}@sfedu.ru

Abstract. This work is aimed at creating tools for automatic parallelization of iterative loops nests on computing systems with distributed memory. The automation for parallelization of the Gauss-Seidel algorithm for the Dirichlet problem on a high-performance cluster is given as an example. To be able to parallelize the original loops nest, we use tiling and hyperplane method (wavefront). Parallelization automation tools are proposed to be built on the basis of a parallelizing system with a high-level internal representation. Tiling and the hyperplane method are performed automatically by the Optimizing Parallelizing System. The results of numerical experiments demonstrating a significant speedup are presented. The relevance of the study is increasing due to the emergence of high-performance "supercomputer-on-a-chip" microprocessors with thousands of cores, which have higher performance than previous multi-core processors. Recommendations are given for creating optimizing parallelizing compilers for such microchips.

Keywords: automatic parallelization · distributed memory · program transformations · data allocation · data transfer

1 Introduction

This paper describes ways to create automatic tools for parallelization of programs for computing systems with distributed memory.

Industrial parallel compilers (GCC, ICC, MS-Compiler, LLVM) parallelize programs for shared memory computing systems. The problems of creating automatic parallelizing compilers are noted in [1]. Many papers describe systems for automatic creation of parallel code for distributed memory computing systems (DMCS), which involve adding pragmas to the text of a sequential program without preliminary transformation [1–4].

For computer systems with distributed memory inter-processor data transfer becomes a time-consuming operation. For high-performance clusters, such transfers can cancel out the speedup from parallelization and even cause slowdowns.

Research supported by Russian Science Foundation grant No. 22-21-00671, https://rscf.ru/project/22-21-00671/.

To effectively map a program onto such computer systems, the program must meet very strict requirements. But recently, multi-core processors, sometimes called "supercomputers on a chip", with tens, hundreds and thousands of cores have appeared. [5–7], Transferring data between processor cores on the same chip takes much less time than on a communication network (Ethernet, Infiniband, PCI-express, etc.). This causes an expansion of the set of efficiently paralleliz-able programs and makes it worthwhile to develop parallelizing compilers for such systems. Automatic mapping of programs to programmable architectures (High Level Synthesis) includes the problems of parallelization on DMCS [8].

[9] describes many problems of linear algebra and mathematical physics, for the parallel solution of which cyclic transfers are used on DMCS.

[1] states: "Data distribution and distribution of computation are closely related - changing data distribution in a simple way often requires a complete rewrite of the computation and communication code.". Block-affine array placements are well suited as a way of distributing data for these types of computations [10,11], since such layouts are described by a small number of parameters and all the most commonly used ways of allocating non-sparse multidimensional arrays are included in the scope of such layouts.

Overlapped Data Distribution Method [12,13] significantly speeds up parallel iterative algorithms by reducing the number of transfers when the sets of transferred data elements become bigger. The number of inter-processor transfers depends on the placement of data in distributed memory. The paper [14] considers the problem of aiding automatic parallelization of a program loop on DMCS and minimizing inter-processor transfers. Loops are parallelized if they contain only assignment statements and use one-dimensional arrays. Works [12,15,16] describe highly efficient transformations of iterative algorithms for solving differential equations of mathematical physics containing the Laplace operator. These transformations used data localization and parallelization for shared memory computing systems. Optimizing transformations in these works were carried out manually.

Automatic tiling for not tight loop nests is described in [17]. Automatic tiling for nests of iterative loops is described in [18]. In [19] automatic parallelization of nests of iterative type loops is given on a GPU. [20,21] describes the implementation of automatic skew tiling and parallelization of the Gauss-Seidel algorithm in OPS (Optimizing parallelizing system) [22] for the Dirichlet problem on a computer system with shared memory and a ten-fold acceleration on an 8-core processor (with shared memory).

This paper presents automatic implementation in OPS for tiling and parallelization of iterative loop nests on DMCS. In this case, tiles must be processed in parallel. Generation of MPI code, inter-processor data transfers and data placement in distributed memory is implemented. The results of numerical experiments demonstrating the speedup are presented. These numerical experiments show that even for high-performance clusters, tools for automatic creation of efficient programs can be useful.

A list of program transformations is given that should be in a compiler that automatically parallelizes a wide variety of nests of iteration-type loops on DMCS. Recommendations are given for the internal representation of a parallelizing compiler for DMCS.

2   Nests of Loops of Iterative Type

We will consider loop nests of a special form, which we will call iterative. These are loops of the kind shown in Listing 1.1. Here we rely on examples of loops that process 2D meshes, but the methods used generalize to more dimensions in a simple way.

Loop nests of this kind appear in implementations of finite difference methods. It is possible to apply the "skewed tiling" transformation to them, which changes the order of loop iterations, but the order of memory accesses does not change, which makes the transformation equivalent. After that, the hyperplane method is applied, in which [21] tiles are executed in parallel.

Listing 1.1. Example of an iterative loop nest

```
for (int k = 0; k < K; ++k ) %// Loop over iterations (1)
      for (int i = 1; i < N - 1; ++i ) // Loop over mesh
          elements (2)
          for (int j = 1; j < M - 1; ++j )
              u[i][j] = a*u[i-1][j] + b*u[i + 1][j] +
              c*u[i][j-1] + d*u[i][j + 1] + e*u[i -
                  1][j - 1] +
              f*u[i + 1][j - 1] + g*u[i + 1][j - 1] +
              h*u[i + 1][j + 1] ;
```

The following restrictions were chosen to represent a wide enough class of programs that would be possible to process automatically:

- coefficients before array entries u (a, b, c, d, e, ...) can be constants, variables, array elements.
- The entire loop nest is tight.
- Values of multidimensional arrays are calculated based on their neighboring elements in the body of the inner loop. Dimension of the arrays (whose values are calculated) is equal to the depth of the loop over the mesh elements (2),
- Array indices are loop counters over mesh elements and they do not depend on the loop counters over iterations (1).
- For each array entry, each index depends on one (only one) of loop counters. For example, $u[i][j] = u[i - 1][j]$ or $u[j][i] = u[j - 1][i]$ (the difference between the corresponding indices is a constant), and the assignment of type $u[j][i] = u[i][j]$ is not allowed.
- Coefficient before loop counter used inside array index is equal to 1 (i.e. each index has the form $[I + C]$) where C is a constant
- Also, there are no *goto, continue, break, exit* statements in the entire nest.

- The loops bounds are arbitrary.
- for "iterative loops" restrictions on accessing arrays are replaced by the following: $X[i1 - a1, i2 - a2, ..., i_n - a_n] = X[i1 - b1, ..., i_n - b_n] + A[...]$, i.e. coefficients before counters $= 1$, each index (dimension) depends on one counter, other arrays can be accessed by similar indices, but elements are assigned always in a single array.

An example of the loop nest and its transformation is shown in Fig. 1. This loop nest is used as an example in this work.

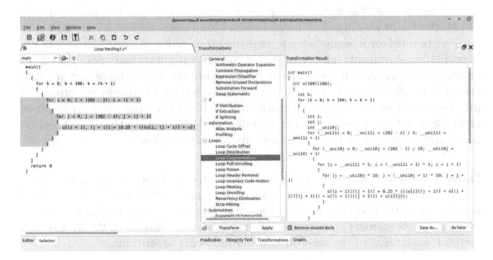

Fig. 1. Applying rectangular tiling to two inner loops in OPS. The result is a nest of loops with dimension of 5

Memory Allocation for Data Transfers. When placing data for the mesh method in distributed memory, it is necessary to take into account the need to transfer data between nodes that process adjacent blocks of the mesh. When using block-affine array placements, the typical way to allocate data in distributed memory would be to partition the mesh into blocks and distribute blocks among nodes.

With such placements, during the iterations of the method, it will be necessary to send elements of block boundaries between neighboring computing nodes. During calculations, the nodes will have to store those elements that were received from their neighbors. Block layouts with overlapping arrays are suitable to provide a generic way to process data for a particular computational method.

Each node will store in its memory not only the elements of the block, but also the boundary elements of neighboring blocks up to a certain depth.

After placing the data on the computing nodes, iterations of the computational method are performed. To do this, each node stores a list of those grid blocks that are in its memory. Block metadata stores information about the information dependencies associated with each block. The order of block calculations within a node is chosen so that the node performs calculations with the blocks for which data has already been prepared. Data transfers are performed after the completion of calculations with the current block if there is a dependency that needs to be satisfied by those transfers.

Each block is identified by coordinates in n-dimensional grid that are calculated from the way original mesh was split into these blocks. We will call the set of these points "block iteration space" to represent different ways of ordering the blocks for processing as in n-dimensional loop nest.

The parallel processing scheme implemented makes it possible to use the hyperplane (wavefront) method for processing computational mesh blocks in distributed memory. This is achieved by changing the order the blocks are processed in to allow several blocks to be processed in parallel. Blocks are processed in separate "waves", each wave consisting of blocks on the same hyperplane of block iteration space. Computing nodes work independently of each other, performing transfers of boundary elements as they are ready.

Data Transfers. In the method used, each iteration requires data transfers. All data elements of the computational mesh are distributed in the form of blocks among the computational nodes, each node has a set of neighboring block elements in its memory. During the execution of one iteration of the hyperplane method, each node works according to the following algorithm:

- Get the next block from the queue.
- If needed, get the border elements of this block from its neighbors.
- Process the taken block.
- Pass border elements of the finished block to other neighbors

Each node stores its blocks in a queue to restrict the order in which the blocks are processed. All nodes have information about which boundary elements are required to process each block, which node stores these elements, and where to send other boundary elements obtained after processing the block.

After the completion of the next iteration, all blocks must update their boundary elements from their neighbors again, so this algorithm is repeated. For illustration, we present the scheme of data exchange between nodes on Fig. 2, which is required for the example loop to work in the same way as the initial loop nest.

The indicated scheme of node interactions can be easily generalized for iterative loop nests of any dimension, given restrictions listed earlier. For example, in the case of the Gauss-Seidel method for solving the three-dimensional Dirichlet problem, all computing nodes can work according to the described algorithm with three-dimensional arrays. In order for the final results to be equivalent to the sequential solution, it is necessary that:

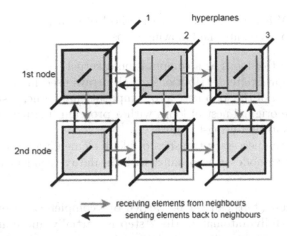

Fig. 2. Element interchange order

- The order in which blocks are processed on each node prevents deadlocks.
- The necessary transfers of boundary elements for each node were indicated in the mesh metadata that is calculated before the main computation. Transfer data is compiled from the information dependencies in the initial loop nest.
- The chosen way of placing blocks on computer nodes is correct given the necessary data transfers.

In the case of a 3D problem, the data transfer directions are similar, but each block may need to send data to three neighboring blocks instead of two and receive data from three others.

Run-Time Library. In order to significantly simplify the task of code generation, it is expedient for the selected parallel processing scheme to be implemented as a "layer" between the created parallel program and low-level data exchange functions provided by the selected parallel processing framework.

All operations of transferring mesh blocks and boundary elements are made in the form of library functions, the role of which is to ensure data storage on the nodes in accordance with the selected processing scheme and the selected data distribution.

The role of the compiler is to automate the generation of a parallel program in a following way:

- Finding suitable loop nests.
- Selecting a parallel processing scheme.
- Filling in data about information dependencies between blocks.
- Conversion of the original loop nest to a form that allows using it as a computational kernel in the selected scheme.
- Generation of a new parallel program based on library function calls, to which the generated computational kernel is passed as a parameter.

Automation. When using the chosen method of creating a parallel program, it is necessary to automate the following steps:

1. Selecting a nest of loops as a computational kernel
2. Determine appropriate distributions for all arrays used in the loop nest
3. Transformation of the loop nest into computational kernel inside a function
4. Replacing the original nest of loops with a program fragment that organizes parallel computation using the library
5. Determining the necessary transfers of block elements for the selected block distribution over nodes in accordance with the data dependencies in the source loop nest

Not all of these steps can yet be carried out completely automatically. The 1, 3, 4 steps are fully automated, the 2 step is partially automated, and the 5 step still needs to be provided with data about the dependencies between the blocks, which in the future will be automatically formed according to the lattice graph of the original loop nest.

The creation of the parallel program presented in this paper was performed partially automatically using OPS. The loop nest is automatically transformed by skew tiling and the hyperplane method. These transformations lead to complex variable index expressions, which are then optimized by the normal OPS compiler transformations. OPS has block-affine allocations of arrays in distributed memory, which are not yet automatically connected in this example. Block-affine placements with overlaps are implemented in OPS, but for one-dimensional loops mapped onto DMCS with a ring communication network. A mesh network may require more complex 2D or 3D layouts with overlaps.

3   Benchmark Results

Benchmark Environment. Performance tests of programs obtained according to the described methods were carried out on two computer clusters:

1. SMRI Blohin Cluster[1]
   - two 20-core Intel Xeon Gold 6230 processors per node
   - more than 500 GB RAM on each node
   - Ethernet communication network
   - up to 3 physical nodes involved
2. IBMX Cluster from Collective High-Performance Computing Center, MMCS SFedU[2]
   - one 3.0 GHz 2-core Intel Xeon 5160 processor with 8 GB RAM per node.
   - up to 13 physical nodes involved
   - DDR Infiniband communication network

---

[1] https://nano.sfedu.ru/en/education/howto/cluster/?CODE=cluster.
[2] http://hpc.sfedu.ru/index.html.

All experiments were carried out with programs obtained from examples of loop nests for two- and three-dimensional Dirchlet problems for the Poisson equation, solved by the Gauss-Seidel method. Parallel programs were obtained using the methods described in the paper. Sequential programs for comparing results (running on a single node) contained only the original loop nests without any extra overhead of data allocation in distributed memory. The number of processes in these results refers to the number of processes running in parallel. When there were not enough physical cluster nodes, some processes ran on the same nodes. All run-time data is an average of 20 runs.

Two-Dimensional Problem Results. The Table 1 shows the achieved speedup of the resulting program compared to sequential implementation. The execution time on bigger mesh is shown in Fig. 3. Only block sizes that make sense for the chosen number of nodes and the method of block distribution were tested. In our case, the number of block rows must be divisible by the number of nodes, otherwise some nodes will not get a single block row. Tested programs executed 10 iterations of the method. Initial data distribution is not taken into account in time measurements, so low number of iterations was sufficient to estimate typical speedup. From the given data, it can be seen that the program generated in this way makes it possible to achieve speedup in comparison with the original sequential program.

Table 1. Speedup for 10 iterations over different node counts and block sizes

	4096			8192		
number of processes	4	8	16	4	8	16
block size						
128	1.10	1.00	1.58	–	–	–
256	1.28	1.52	7.19	1.04	1.42	1.96
512	1.54	3.50	–	1.14	1.89	8.07
1024	2.40	–	–	1.75	4.38	–
2048	–	–	–	2.01	–	–

It is clearly seen that the block size and the number of nodes greatly affect the computation time, but the resulting speedup shows the effectiveness of this method. To achieve maximum computational efficiency, it is necessary that each node processes the smallest possible number of blocks, because the overhead of organizing data storage and exchange is very noticeable.

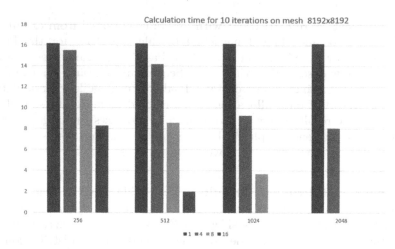

Fig. 3. Time of the computational method at 10 iterations on a grid of size 8192 (in seconds)

From the above data, we can conclude that the method of storing data and transferring it creates large overhead costs, the impact of which decreases with an increase in the number of iterations. At the same time, if each node stores the smallest possible number of blocks, then it is possible to achieve a noticeable speedup compared to running on one node, when the entire mesh is stored in one array.

3d Problem Benchmark Results. Experiments were also carried out on the processing of a three-dimensional loop nests, which solves a similar problem. The running time of the program on a lattice of size 512 with blocks of size 128 on an IBMX cluster is shown in Table. 2:

Table 2. Time to complete 10 iterations of a 3D problem on a 512 mesh (in seconds)

Number of nodes	time, sec.
1	20.41
4	7.27
8	6.30
16	3.77

As can be seen from this data, the overhead of data transfers greatly reduces parallel speedup. But this problem can be solved by optimizing the run-time library functions. It is possible to optimize block storage on each node. It is also possible to use different node interaction schemes, for example, to implement a job queue scheme so that the blocks are selected for processing to better balance the load on the nodes. Load imbalance could arise because of hyperplane method (wavefront) applied to a small number of parallel processes.

4 Automation of Mapping Programs onto DMCS

On the basis of the data presented in this paper, we can conclude that an optimizing compiler that effectively maps iterative loop nests onto DMCS should include the following program transformations.

1. Selecting a loop nest as a computing kernel.
2. Checking if the loop nest is of iterative type? Determining the parameters of this nest: the dimension of the calculated array; constants in index expressions of a computed array.
3. Skewed tiling.
4. Hyperplane method, which will allow simultaneous execution of skewed tiles.
5. Prediction of optimal tile sizes, for example based on the static analysis to approximate amount of computations performed on average per each data element.
6. Block-affine placements of multidimensional arrays with overlaps in accordance with the capabilities of the communication network. With a ring overlap network for two neighboring nodes, for a two-dimensional mesh - for 4 neighboring nodes, for a three-dimensional mesh - for 6 neighboring nodes.
7. Inserting the necessary data transfers into the text of the program with the selected placement of blocks on the nodes in accordance with the informational dependencies in the source loop nest.
8. Code generator for parallel execution on a DMCS (MPI or SHMEM).
9. Loop nesting to match the number of iterations of the loop being parallelized and the number of DMCS nodes
10. Transformations that optimize index expressions of arrays after they have been complicated by tiling, hyperplanes, and nesting.

Some of the above transformations are fully automated based on OPS. The above list of transformations and recommendations for the internal representation can be considered as the basis of a project to create a compiling parallelizing system oriented to new generation of distributed memory computing systems.

5 Conclusions

When making the creation of a parallel code for DMCS more automatic, some difficulties arise. It is advantageous to overcome these problems in a translator from C to C, for example, based on the Optimizing Parallelizing System.

This work continues the research of the authors on the creation of optimizing parallelizing compilers for DMCS. Previous works have discussed automatic parallelization of simple program loops that contain only occurrences of one-dimensional arrays and no other loops.

This work demonstrates the automation of parallelization of nests of iterative loops, which require large amounts of calculations and are found in many problems of mathematical modeling.

One of the results of this work is a list of transformations that need to be included in the DMCS compiler, and recommendations for the internal representation of such a compiler. Arguments are made to support a high level internal representation.

The obtained speedups of the program under consideration demonstrate the prospects of developing compilers for DMCS.

The study was supported by the Russian Science Foundation grant No. 22-21-00671, https://rscf.ru/project/22-21-00671/.

References

1. Bondhugula, U.: Automatic distributed-memory parallelization and code generation using the polyhedral framework. In: Technical report, ISc-CSA-TR-2011-3, p. 10, September 2011. https://mcl.csa.iisc.ac.in/downloads/publications/uday11distmem-tr.pdf
2. DVM-system for parallel program development — DVM-system. https://dvm-system.org/ru/about/
3. Kataev, N., Kolganov, A.: Additional parallelization of existing MPI programs using SAPFOR. In: Malyshkin, V. (ed.) PaCT 2021. LNCS, vol. 12942, pp. 41–52. Springer, Cham (2021). https://doi.org/10.1007/978-3-030-86359-3_4 ISSN: 1087-4089
4. Kwon, D., Han, S., Kim, H.: MPI backend for an automatic parallelizing compiler. In: Proceedings Fourth International Symposium on Parallel Architectures, Algorithms, and Networks (I-SPAN 1999), pp. 152–157, June 1999. https://doi.org/10.1109/ISPAN.1999.778932. ISSN 1087-4089
5. Processor from NTC "modul". https://www.cnews.ru/news/top/2019-03-06_svet_uvidel_moshchnejshij_rossijskij_nejroprotsessor
6. SoC esperanto. https://www.esperanto.ai/
7. Peckham, O.: SambaNova launches second-gen DataScale system. https://www.hpcwire.com/2022/09/14/sambanova-launches-second-gen-datascale-system/
8. Dordopulo, A.I., Levin, I.I., Gudkov, V.A., Gulenok, A.A.: High-level synthesis of scalable solutions from C-programs for reconfigurable computer systems. In: Malyshkin, V. (ed.) PaCT 2021. LNCS, vol. 12942, pp. 88–102. Springer, Cham (2021). https://doi.org/10.1007/978-3-030-86359-3_7
9. Prangishvili, I.V., Vilenkin, S.Ia., Medvedev, I.L.: Parallelnye vychislitelnye sistemy s obshchim upravleniem. Energoatomizdat, Moskva (1983). https://www.livelib.ru/book/1000878401-parallelnye-vychislitelnye-sistemy-s-obschim-upravleniem-iveri-prangishvili
10. Shteinberg, B.Ia.: Blochno-affinnye razmeshcheniia dannykh v parallelnoi pamiati. Informatsionnye tekhnologii 6, 36–41 (2010). https://www.elibrary.ru/item.asp?id=14998775. ISSN 1684–6400. Place: Moskva Publisher: OOO "Izdatelstvo Novye tekhnologii"
11. Shteinberg, B.Ia.: Optimizatsiia razmeshcheniia dannykh v parallelnoi pamiati. Prioritetnye natsionalnye proekty. Obrazovanie. Izdatelstvovo Iuzhnogo Federalnogo Universiteta, Rostov-na-Donu (2010). ISBN 978-5-9275-0687-3
12. Ammaev, S.G., Gervich, L.R., Steinberg, B.Y.: Combining parallelization with overlaps and optimization of cache memory usage. In: International Conference on Parallel Computing Technologies, pp. 257–264 (2017)

13. Gervich, L.R., Steinberg, B.Ya.: Automation of the application of data distribution with overlapping in distributed memory. Bulletin of the South Ural State University. Ser. Math. Model. Program. Comput. Softw. (Bull. SUSU MMCS) **16**(1), 59–68 (2023)
14. Krivosheev, N.M., Steinberg, B.Y.: Algorithm for searching minimum inter-node data transfers. In: Procedia Computer Science, 10th International Young Scientist Conference on Computational Science. Accessed 1 July 2021
15. Levchenko, V., Perepelkina, A., Zakirov, A.: DiamondTorre algorithm for high-performance wave modeling **4**(3), 29. https://doi.org/10.3390/computation4030029. https://www.mdpi.com/2079-3197/4/3/29. ISSN 2079–3197
16. Perepelkina, A.Y., Levchenko, V.D.: The DiamondCandy algorithm for maximum performance vectorized cross-stencil computation (225), 1–23. https://doi.org/10.20948/prepr-2018-225-e. https://keldysh.ru/papers/2018/prep2018_225_eng.pdf. ISSN 20712898, 20712901
17. Song, Y., Li, Z.: A compiler framework for tiling imperfectly-nested loops. In: Carter, L., Ferrante, J. (eds.) LCPC 1999. LNCS, vol. 1863, pp. 185–200. Springer, Heidelberg (2000). https://doi.org/10.1007/3-540-44905-1_12
18. Song, Y., Li, Z.: Automatic tiling of iterative stencil loops. In: Carter, L., Ferrante, J. (eds.) LCPC 1999. LNCS, vol. 1863, pp. 185–200. Springer, Heidelberg (1999). https://doi.org/10.1007/3-540-44905-1
19. Christen, M., Schenk, O., Burkhart, H.: PATUS: a code generation and autotuning framework for parallel iterative stencil computations on modern microarchitectures. In: 2011 IEEE International Parallel & Distributed Processing Symposium, pp. 676–687 (2011)
20. Steinberg, B.Ya., Steinberg, O.B., Oganesyan, P.A., Vasilenko, A.A., Veselovskiy Null, V.V., Zhivykh, N.A.: Fast solvers for systems of linear equations with block-band matrices. East Asian J. Appl. Math. **13**(1), 47–58 (2023). https://doi.org/10.4208/eajam.300921.210522. https://global-sci.org/intro/article_detail/eajam/21301.html. ISSN 2079–7362, 2079–7370
21. Vasilenko, A., Veselovskiy, V., Metelitsa, E., Zhivykh, N., Steinberg, B., Steinberg, O.: Precompiler for the ACELAN-COMPOS package solvers. In: Malyshkin, V. (ed.) PaCT 2021. LNCS, vol. 12942, pp. 103–116. Springer, Cham (2021). https://doi.org/10.1007/978-3-030-86359-3_8
22. Optimizing parallelizing system (2018). https://www.ops.rsu.ru

Didal: Distributed Data Library for Development of Parallel Fragmented Programs

Victor Malyshkin[1,2,3] and Georgy Schukin[1,2(✉)]

[1] Institute of Computational Mathematics and Mathematical Geophysics,
SB RAS, Novosibirsk, Russia
{malysh,schukin}@ssd.sscc.ru
[2] Novosibirsk State Technical University, Novosibirsk, Russia
[3] Novosibirsk State University, Novosibirsk, Russia

Abstract. Nowadays with rapid evolution of high-performance computing systems it's becoming essential to have tools to simplify development of efficient portable parallel programs for these systems. Fragmented programming is a technology where parallel program is represented as a collection of pieces of data (data fragments) and computations on these pieces (computation fragments), able to be tuned to the resources of a computing system and automatically provide such facilities as dynamic load balancing. Didal is a distributed data library to support development of efficient parallel fragmented programs on distributed memory supercomputers. The library contains facilities for data partitioning, distribution and load balancing. In this paper foundations of the library are explained and applicability of the library is demonstrated with Particle-in-Cell (PIC) method implementation, which shows performance comparable to conventional parallel programming tools.

Keywords: Parallel programming · Fragmented programming · Distributed data · High-performance computing

1 Introduction

High-performance computing systems of today consist of hundreds of computing nodes and thousands of cores. To utilize all these resources a way to develop efficient and portable parallel programs for distributed memory is required. Conventional parallel programming tools for distributed memory - such as MPI - usually are too low-level and in many cases are not sufficient to create portable and easily re-configurable parallel programs. Didal is a library aimed to simplify creation of efficient portable parallel programs. As a base model for parallel program fragmented programming model is used.

This work was carried out under state contract with ICMMG SB RAS 0251-2022-0005.

V. Malyshkin (Ed.): PaCT 2023, LNCS 14098, pp. 30–41, 2023.
https://doi.org/10.1007/978-3-031-41673-6_3

Fragmented programming [1] is a technology where parallel program is represented as a collection of pieces of data (data fragments) and computations on these pieces (computation fragments), able to be tuned to the resources of the system and automatically provide such facilities as dynamic load balancing.

Didal is a distributed data library to support development of efficient parallel fragmented programs on distributed memory supercomputers. The library contains facilities for data partitioning, distribution and load balancing. In this paper foundations of the library are explained and testing of its performance is presented.

Fragmented programming technology is a part of the active knowledge approach [2]. This approach is aimed at automation of construction of efficient parallel programs in order to reduce complexity and laboriousness of programs development and allow programmers to focus on the subject domain, not on low-level problems of parallel programming.

It is known that implementation of such automation in general case is an algorithmically hard problem, so no general solution is to be expected here. Nevertheless, in many limited subject domains such automation can be achieved if the subject domain has a lot of particular solutions already exist.

The idea is to *automatically* reuse existing software accumulated in particular subject domains, rather than *manually* reuse of the software in form of, say, a subroutines library. To achieve this goal the active knowledge approach suggests making a formal description of existing pieces of software in a subject domain. Such description must allow automatic search and application of the pieces for solution of a problem, formulated in terms of the subject domain. For that a theoretical basis was proposed in [3]. Although main ideas were formulated there, application of the approach in particular subject domains still requires research.

Of interest are ways to incorporate particular algorithms, programs and other solutions, specific to a subject domain. In present paper we study a case of incorporating some high performance numerical programs development practices, methods and algorithms for particular parallel programs class within the framework of fragmented programming technology. In particular, we propose particular algorithms implemented as the Didal library to make the algorithms be available for automatic reuse for parallel programs construction. In the paper, however, these algorithms implementations are employed manually to ensure their efficiency with real applications.

The paper is organized as follows. The Sect. 2 contains review of related works. The Sect. 3 describes fragmented programming technology. The Sect. 4 describes Didal's architecture and design decisions. The Sect. 5 contains description of PIC method application used to test Didal's applicability and Sect. 6 contains results of performance experiments.

2   Related Works

There is a vast body of research in the field of high-level parallel programming, with many projects, as active as inactive ones. Such systems as Charm++ [4],

Chapel [5] or Legion [6] represent complete frameworks or languages for distributed memory parallel programming. While these systems allow for high-level creation of parallel programs, they may contain some downsides. First, an user of a system may be required to learn completely new language or framework. Second, usually these systems manage not only data but also computations, and thus it may be not easy to combine a program in such system with another parallel programming tool or paradigm or to use some third-party library. Thus, it was decided to make Didal a library in conventional C++ programming language which provides distributed data facilities, so a programmer controls how data is processed, be it in a single thread or using OpenMP, multiple threads, GPU, etc.

Other systems for parallel programming may use a form of language extension, such as mpC [7] or DVM [8]. These systems require special compiler which understands the extension, whereas Didal, being a library, works with any C++ compiler. Also, the base language they extend (C or FORTRAN) may lack some desired capabilities; in Didal we may use C++ static and dynamic polymorphism to enable compile-time optimizations or make resulting algorithms and data types extensible and adaptable.

Examples of libraries which provide facilities for distributed data (distributed data containers) are DASH [9], BCL [10], STAPL [11], HCP++ [12] and UPC++ [13]. These libraries usually hide data partitioning details from an user and provide imitation of a global address space (PGAS, partitioned global address space). In such libraries a unit of access is a single data element; when a remote element is accessed, it causes a separate remote operation. Although this allows to simplify parallel programs, many accesses to remote elements can degrade efficiency due to high latency. Didal supports explicit fragmentation where data elements can be grouped into data fragments and be accessed with a single operation, also making a place of such remote access explicit.

3   Fragmented Programming Technology

Fragmented programming represents a way to construct a parallel program which then can be executed on a wide variety of computing machines and for which different optimizations can be automatically employed.

Fragmented program is a collection of pieces of data (data fragments) and computations (computational fragments) which compute output data fragments from input data fragments. Fragmented program can be executed in parallel by executing independent (i.e. without data dependencies) computational fragments in parallel. In a case of distributed memory, distribution of data and computational fragments on different nodes of a supercomputer may be required.

Among the benefits of a fragmented program is that the same program can be used for many different architectures, provided that a special run-time system or library exists which performs mapping of the program to existing resources. Many optimization strategies, such as dynamic load balancing, can be automatically performed by changing distribution of data/computational fragments, without much intervention from a programmer. To optimize program's performance for a particular machine, strategies for data partitioning, distribution, load balancing, etc. can be changed (by run-time system or a programmer), possibly in automatic or semi-automatic way.

4   Didal: Distributed Data Library

Didal is a C++ library designed to support fragmented program development for distributed memory machines. The library presents high-level abstractions and interfaces which should simplify creation of portable efficient parallel fragmented programs. Leveraging capabilities of C++ static and dynamic polymorphism, the library can also serve as a platform for development of reusable fragmented algorithms and distributed data types.

The main unit in the library is a distributed collection of objects. A data fragment is an example of such object. The collection is distributed, so each computing node stores some objects from the collection. Each object has an unique global identifier, by which it can be accessed from any node. The library allows many operations with distributed objects: creation, copying, moving, deletion, etc. Objects' distribution, location and synchronization of access are managed by the library internally.

Fragmented programs in Didal are written in terms of working with distributed collections of objects (data fragments).

4.1   Library's Structure

The structure of the library is presented in Fig. 1, where each layer utilizes facilities provided by lower layer.

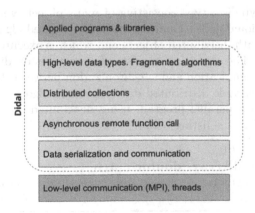

Fig. 1. Didal's structure.

The core of Didal is distributed collections of objects. This layer also includes different data partitioning, migrating and balancing strategies for these collections. Using these distributed collections, high-level distributed data types and parallel algorithms can be constructed and it's with these data types and algorithms an user of the library will be interfacing. Domain-specific libraries and applied programs can be created using Didal (top level on the picture, outside of Didal's scope).

Distributed collections itself are functioning using asynchronous remote function call as the main way of communication between computing nodes of a distributed memory machine. This mechanism allows to call any function/method remotely on any node from any node and receive the result asynchronously.

To make possible remote function call with any types of arguments and return value, data needs to be transmitted through network, first arguments to the calling site and then return value back again. This requires an ability to send/receive data (communication) as well to serialize/deserialize it to make it suitable for any data type to be transmitted using low-level primitive communication routines (serialization). Didal automatically provides serialization for built-in elementary types, simple types whose objects are stored as one memory block, as well as standard C++ containers such as vector, list or map. Serialization of objects of any other complex type may be easily build by recursively serializing its components.

The bottom layer (also outside of Didal's scope) is a system layer and contains such stuff as low-level communication primitives (for example, MPI), thread and memory management, etc. This layer is provided by standard C++ library and an operating system.

The next subsections describe Didal's components in more details.

4.2   Distributed Data Types and Algorithms

Distributed data type is a type consisting of many objects which are stored on different computational nodes simultaneously. Distributed algorithm is an algorithm that works with distributed objects and which is executed simultaneously by different nodes. The algorithm consists of operations on distributed objects with necessary communication and synchronization between them. Didal allows to build different reusable distributed data types and algorithms.

The example of a parallel program using distributed data types and algorithms is presented in the listing below:

```
const int N = 100;
ddl :: DistributedArray<double> a(N);
ddl :: initRandom(a); // init distributed array
double sum = ddl :: sum(a); // compute sum of elements
```

4.3 Distributed Collections

Distributed collection of objects is a base for any distributed data type in Didal. A collection works as an associative array (dictionary, map), where each objects has its unique global identifier. Objects are stored distributively on different computing nodes; usually each node stores a subset of objects from the collection. Different distribution strategies can be used to control how objects are actually distributed. A location strategy is used to locate any object (i.e. to determine its storing node) from any node.

A distributed collection supports such operations on its objects as creation, removal, copy, modification, etc. (each can be done with any object from any node, a local or a remote one). An object's id is used to identify the object globally.

Listing below presents an example of working with a distributed collection:

```
const int N = 100, numOfBlocks = 10;
// Partition array onto blocks and distribute them.
ddl::StaticBlockDistribution<1> distr(N, numOfBlocks);
ddl::DistributedCollection<int,
    Block1D<double>> coll(&distr);
// Each process creates data blocks assigned to it.
for (auto ind: distr.localIndices()) {
    auto block = randomBlock(distr.blockSize(ind));
    coll.add(ind, block);
}
// Different ways to operate on blocks.
double sum = 0, sum2 = 0;
for (auto ind: distr.allIndices()) {
    // Request block and process it.
    auto blockFuture = coll.get(ind);
    auto block = blockFuture.get(); // wait for block
    sum += sumElements(block);
    // Perform operation on block and get the result.
    auto callFuture = coll.call(ind, &sumElements);
    sum2 += callFuture.get(); // wait for result
}
```

4.4 Asynchronous Remote Function Call

The main way of accessing a remote data (object) in Didal is a remote function call. With this call one can get a copy of the remote object (for read) or perform some operation on the object and receive a result. Didal also allows to call procedures on remote objects (without returning result) or to call class methods on these objects if their data type is a class.

Remote call is asynchronous which means that a caller can initiate a remote operation and then try to access its result later. This allows to overlap computations and communications (a caller can perform another actions while remote operation is being completed) and to start several (independent) operations at once to minimize communication latency. Standard C++ *future* mechanism is used for result's retrieval.

In this scheme only caller initiates a communication, so receiving (remote) site needs not to be aware. This allows for greater flexibility for programming communications, but may require synchronization when remote operation modifies data.

4.5   Synchronization

When calling an operation on remote object, a caller needs to be sure that this object exists and, in case of mutable objects, has a correct state. Due to distributed nature of a typical parallel Didal application, there could be time differences between access to a remote object and actual creation/update of this object. Synchronization is necessary to ensure that a correct value of the remote object is accessed. Didal supports several modes of synchronization in its distributed collections.

Any object is accessed by its id, so when a remote node is checked for an object with some id, it will wait until an object with this id is actually created. If only immutable (non-modifiable) objects are used in a program, this by-a-fact-of-creation synchronization is usually enough to ensure program's correctness.

When an object can be modified, things become more complex, because now we need to be sure that a remote object is in correct state when accessed. Particular details of synchronization may depend on an actual program and problem at hand, but for this case Didal supports additional synchronization with epochs. Each object can additionally store its corresponding epoch - usually a single integer number; when the object is updated, its epoch number increases. Calling site, when accessing an object with a remote function call, also specifies its desired epoch number. The object will be accessed only if its current epoch number is no less than this specified epoch number, in another case the access will be postponed.

4.6   Serialization and Communication

For any remote function call its arguments need to be transmitted to the remote node and the result needs to be transmitted back. For actual data transmission in Didal low-level communication utilities (such as MPI) are used. These utilities usually work with data as with an array of bytes. To be able to transmit objects of any data type serialization is used which first converts any object to an array of bytes on the sending node and then restores it on the receiving node. Serialization allows to transfer any complex data types, such as containing pointers and references.

Low-level communication between nodes in Didal is done via asynchronous messages. Each message has a body (actual data to transmit) and a header (contains such information as receiver id for message routing). On each node usually there is a separate thread for processing of incoming messages.

5   Program Example: Particle-In-Cell (PIC) Method

Particle-in-cell (PIC) method is used in many areas of computational physics, for example to simulate behaviour of dust or plasma. In the current example we are using PIC to simulate dust clouds under gravitational forces [14].

Simulation domain contains particles and regular grids which values for density distribution and gravitational forces and potential. These grids form a mesh of regular cells, with values in a cell's vertices and with particles inside a cell. Usually, only particles inside a cell affect field values for this cell.

The simulation process consist from the following repeating steps:

1. Density distribution is computed based on particles' current position
2. Gravitation potential is computed by solving Poisson's equation on the grid
3. Gravitational forces are computed on the grid from potential
4. Particles' position and velocity are updated based on computed forces

Parallelization of the method is performed by dividing cells and particles between processes/nodes and performing computation on different cells in parallel. For locality reasons particles in a cell are stored together with this cell. When a particle's position is changed, it can start belonging to another cell, and should be physically transferred to this cell. For distributed memory that means that particles are transferred between computing nodes.

6   Experiments

In the experiments we compared performance of Didal and pure MPI implementations of the PIC method. Both versions utilized the same low-level computational code. For data transfer Didal version used asynchronous access to distributed objects with remote function calls, whereas MPI version used asynchronous MPI calls.

The following hardware and software was used for the experiments:

- MVS-10P OP2 cluster: two Intel Xeon Gold 6248R 24-core processors (48 cores total) per computing node, Intel Omni-Path interconnect
- MPI library: OpenMPI 4.1.3
- Compiler: GCC 10.2.0

Both MPI and Didal versions allowed up to 24 processes per computing node, so each thread in a process ran on a separate core. In Didal version each process used two threads (the main thread with computations and background thread for receiving and processing of messages). MPI version was a single-threaded application and in theory might have used more then 24 processes per node, but in practice its performance degraded in this case. This also allowed to make both versions use shared memory or inter-node network communication with the same number of processes.

For the PIC method its parameters were: 3-dimensional regular mesh consisted of 128^3 cells (128 cells by each dimension) and contained 10^7 particles, 1000 time steps (iterations) were performed. For fragmented (Didal) version the mesh was partitioned onto 8^3 regular blocks (8 blocks by each dimension).

6.1 Results

The first test was a strong scalability test, the size of the problem (number of cells and particles) was fixed. Particles were initially located in a cloud in a center of the domain. For MPI version the mesh was partitioned (by all three dimensions) onto blocks of cells and each process was assigned a single block; all processes were arranged into a three-dimensional lattice topology. For Didal version mesh of fragments was partitioned onto blocks (groups) of fragments and each process was assigned a block of fragments.

Results of working time (measured in seconds) and parallelization efficiency (measured in percents) are presented in Tables 1 and 2 correspondingly; N denotes number of processes. As can be seen from the results, Didal version outperforms MPI version up to 128 processes. Explanation for the speed-up can be fragmentation effect, where smaller fragment size of Didal version allows to utilize cache more efficiently, hence providing faster computing time. For larger number of processes amount of data per process becomes smaller and communication starts to play major role. Utilizing nearest-neighbor asynchronous communications and mostly avoiding global synchronization allowed Didal version to show more-or-less comparable scalability with MPI version, although the gap increases for a large number of processes.

Explanation for this gap can be follows. First, current version of Didal used MPI internally on low level with multi-threaded MPI mode, whereas pure MPI program used single-threaded MPI mode. Single-threaded MPI mode is usually more efficient than multi-threaded. Second, to be able to communicate data of any type, Didal uses data serialization. This serialization introduces additional overhead. Optimization of communications is a topic of further research.

Table 1. Strong scalability, working time.

N	1	2	4	8	16	32	64	128	256	512
MPI	1900.11	961.07	482.03	238.08	172.15	122.66	85.49	37.35	16.89	8.31
Didal	996.91	522.04	280	161.91	123.53	91.64	65.44	36.37	22.73	11.49

Table 2. Strong scalability, efficiency.

N	2	4	8	16	32	64	128	256	512
MPI	98.9	98.6	99.8	68.9	48.4	34.7	39.8	43.9	44.7
Didal	95.5	89	76.9	50.4	33.9	23.8	21.4	17.1	16.9

Table 3 shows parallelization efficiency for separate steps of the PIC application: computation of density distribution (Density), gravitational potential (Potential), gravitational forces (Forces), update of particles' position and velocity (Part. update) and migration of particles between cells (Part. migrate).

Update of particles and computation of forces show the best efficiency due to the fact that they doesn't require any communication and are easily parallelizible. Update of particles for MPI version even shows super-linear speedup, which is caused by efficient utilization of cache on a large number of processes, when the size of a cell block per process becomes rather small. For the remaining three steps efficiency gradually diminishes. Computation of potential requires global reduce operation to solve Poisson equation with iterative method. Amount of computations itself is small. Low efficiency for density computation and migration of particles is due to non-uniform distribution of particles and caused by it load imbalance.

Table 3. Strong scaling, steps efficiency.

N	2	4	8	16	32	64	128	256	512
Density (Didal)	95.1	91.4	88.3	56.9	35.3	29.2	24.5	16.1	16.3
Density (MPI)	100.2	109.9	128.4	81.5	61.7	42.7	32	28.6	27
Potential (Didal)	93.1	91.1	72.8	53.2	39.4	14.7	9.7	5.8	5.3
Potential (MPI)	95.9	91.3	86.1	73.5	60.4	61.3	17.5	9.9	5.3
Forces (Didal)	96.9	95.8	92	85.7	88.9	89.8	89.5	87.7	85.3
Forces (MPI)	99.1	94	86.2	82.1	79.5	76.4	77.5	78.5	74.5
Part. update (Didal)	97.2	95.3	93.5	91.1	92.5	92.5	93.4	91.7	95.3
Part. update (MPI)	99.8	100.7	96.7	105.7	125.2	149	170.1	176	182.2
Part. migrate (Didal)	93	75	49.8	23.5	13.7	8.7	8.4	7.8	7.7
Part. migrate (MPI)	97	90.1	91	43.6	24.3	15.8	21.9	29.4	36.3

The second test was weak scalability test, where problem size per process was fixed and total problem size was scaled with a number of processes. Each process contained 128^3 cells with 10^7 particles, which were distributed uniformly in the domain. For fragmented Didal version a cell block on each process was partitioned into 8^3 fragments.

Working time results are presented in the Table 4, and efficiency - in the Table 5. In this test Didal always outperforms MPI version, again due to benefits from fragmentation and asynchronous access to remote data. Now when relation between computations and communications doesn't change, parallel efficiency of both versions is quite close.

Table 4. Weak scaling, working time.

N	1	2	4	8	16	32	64	128	256	512
MPI	294.51	346.97	372.95	393.37	430.48	445.05	493.34	496.68	500.20	601.55
Didal	200.33	222.25	238.65	248.31	277.42	308.57	341.3	354.49	392.81	501.88

Table 5. Weak scaling, efficiency.

N	2	4	8	16	32	64	128	256	512
MPI	84.9	79	74.9	68.4	66.2	59.7	59.3	58.9	49
Didal	90.1	83.9	80.7	72.2	64.9	58.7	56.5	51	39.9

Efficiency for different steps is presented in the Table 6. Again, update of particles and computation of forces shows the best efficiency. Computation of density and potential and migration of particles have better efficiency than in the strong scalability test. Even with uniform distribution of particles efficiency for these phases still drops in both versions, but now because amount of transmitted data per process remains more or less constant, so total amount of transmitted data increases with increasing number of processes, so performance of the network may be not enough. Another reason for decrease in efficiency is that although initial particle distribution is uniform, it may become non-uniform due to particles' migration and cause load imbalance. In that case dynamic load balancing may be required, which is a topic of the future research.

Table 6. Weak scaling, steps efficiency.

N	2	4	8	16	32	64	128	256	512
Density (Didal)	84.4	73.1	69	62.5	54.3	47.8	49.1	46.1	39.8
Density (MPI)	75.7	64.1	58	51.8	49.9	45.2	44.7	44.3	35.2
Potential (Didal)	85.2	80.8	76.7	67.1	59.2	55.1	49.7	41.4	26.5
Potential (MPI)	97.8	93.1	89.1	83.5	82.5	74.9	55.2	41.5	19.9
Forces (Didal)	98	95.4	90.8	78.9	79.5	79.8	79.8	76.7	76.6
Forces (MPI)	98.1	93.5	91.1	82.1	82.1	81	83	80.6	80.2
Part. update (Didal)	101.2	101.6	106.4	106.3	108.5	110.6	108.5	107.7	109.7
Part. update (MPI)	84.4	77.9	75.1	77.5	80	80.9	82.3	87	97.5
Part. migrate (Didal)	95.2	90.8	78.2	55.4	44.8	32.7	28.6	24.9	19.4
Part. migrate (MPI)	86.4	91.8	87.2	56.7	47.4	34.9	37.5	40	36.9

7   Conclusion

The distributed data library Didal was developed to simplify development of parallel fragmented programs for distributed memory supercomputers. It's applicability is tested by PIC method implementation, where it showed comparable or even better performance and not much worse scalability than conventional low-level parallel programming tools such as MPI. Further research includes optimization of the library and testing of different data distribution and dynamic load balancing strategies.

References

1. Kireev, S., Malyshkin, V.: Fragmentation of numerical algorithms for parallel subroutines library. J. Supercomput. **57**, 161–171 (2011). https://doi.org/10.1007/s11227-010-0385-3
2. Malyshkin, V.: Active knowledge, LuNA and literacy for oncoming centuries. LNCS **9465**, 292–303 (2015)
3. Valkovsky, V.A., Malyshkin, V.E.: Synthesis of parallel programs and systems on the basis of computational models, Nauka, Novosibirsk, pp. 128 (1988). In Russian: Valkovsky, V.A., Malyshkin, V.E. (eds.) Sintez parallelnykh program i system na vychislitelnykh modelyah. Nauka, Novosibirsk, 128 str. (1988)
4. Acun, B., et al.: Parallel programming with migratable objects: Charm++ in practice. In: SC '14: International Conference for High Performance Computing, Networking, Storage and Analysis, pp. 647–658. IEEE Press (2014). https://doi.org/10.1109/SC.2014.58
5. Chamberlain, B.L., Callahan, D., Zima, H.P.: Parallel programmability and the chapel language. Int. J. High Perform. Comput. Appl. **21**(3), 291–312 (2007). https://doi.org/10.1177/1094342007078442
6. Bauer, M., Treichler, S., Slaughter, E., Aiken, A.: Legion: expressing locality and independence with logical regions. In: SC '12: International Conference on High Performance Computing, Networking, Storage and Analysis, pp. 1–11. IEEE Computer Society Press, Washington, DC (2012). https://doi.org/10.1109/SC.2012.71
7. Lastovetsky, A.: Adaptive parallel computing on heterogeneous networks with mpC. J. Parallel Comput. **28**(10), 1369–1407 (2002). https://doi.org/10.1016/S0167-8191(02)00159-X
8. Bakhtin, V.A., Krukov, V.A.: DVM-approach to the automation of the development of parallel programs for clusters. J. Program. Comput. Softw. **45**, 121–132 (2019). https://doi.org/10.1134/S0361768819030034
9. Fürlinger, K., Fuchs, T., Kowalewski, R.: DASH: A C++ PGAS library for distributed data structures and parallel algorithms. In: IEEE 18th International Conference on High Performance Computing and Communications (HPCC). IEEE Press (2016). https://doi.org/10.1109/HPCC-SmartCity-DSS.2016.0140
10. Brock, B., Buluç, A., Yelick, K.: BCL: a cross-platform distributed data structures library. In: ICPP '19: 48th International Conference on Parallel Processing, pp. 1–10. Association for Computing Machinery Press, New York (2019). https://doi.org/10.1145/3337821.3337912
11. Buss, A., et al.: STAPL: standard template adaptive parallel library. In: SYSTOR '10: 3rd Annual Haifa Experimental Systems Conference, pp. 1–10. Association for Computing Machinery Press, New York (2010). https://doi.org/10.1145/1815695.1815713
12. Beckman, P.H., Gannon, D., Johnson, E.: HPC++: experiments with the parallel standard template library. In: ICS '97: 11th International Conference on Supercomputing, pp. 124–131. Association for Computing Machinery Press, New York (1997). https://doi.org/10.1145/263580.263614
13. Bachan, J., et al.: UPC++: a high-performance communication framework for asynchronous computation. In: 2019 IEEE International Parallel and Distributed Processing Symposium (IPDPS), pp. 963–973. IEEE Computer Society Press (2019). https://doi.org/10.1109/IPDPS.2019.00104
14. Kireev, S.: A parallel 3D code for simulation of self-gravitating gas-dust systems. In: Malyshkin, V. (ed.) PaCT 2009. LNCS, vol. 5698, pp. 406–413. Springer, Heidelberg (2009). https://doi.org/10.1007/978-3-642-03275-2_40

Trace Balancing Technique for Trace Playback in LuNA System

Victor Malyshkin[1,2,3], Vladislav Perepelkin[1,2,3]([✉]), and Artem Lyamin[2]

[1] Institute of Computational Mathematics and Mathematical Geophysics,
SB RAS, 630090 Novosibirsk, Russia
{malysh,perepelkin}@ssd.sscc.ru
[2] Novosibirsk State University, 630090 Novosibirsk, Russia
[3] Novosibirsk State Technical University, 630073 Novosibirsk, Russia

Abstract. In the paper an improved trace playback technique is presented. Run-time systems are widely used in parallel programming to provide dynamic properties of programs execution. However, run-time system often cause significant overhead. Trace playback is a technique, oriented to improve parallel program execution by reducing the overhead. It consists in recording a special log (called *trace*) while run-time system executes a program. The trace contains enough information on exact actions performed to reproduce the execution without the run-time system. Run-time system overhead is thus eliminated. The technique is usable in such systems as LuNA. The proposed improvement of the technique consists in modification of ("balancing") the trace before trace playback in order to fit more efficiently into given multicomputer. Particular balancing algorithm, as well as experimental study results are presented in the paper. The improvement showed a significant performance increase.

Keywords: LuNA System · Fragmented Programming · Trace Playback · Trace Balancing · Parallel Programs Construction Automation

1 Introduction

Development of parallel programs for numerical simulations on supercomputers is troublesome and laboriousness. This is caused by the fact that provision of satisfactory efficiency of parallel program execution requires solution of diverse problems, related to system parallel programming, such as data and computations decomposition, handling concurrent execution, synchronizing access to shared resources, scheduling computations and communications, etc. Often dynamic memory management, dynamic load balancing, check-pointing and other dynamic properties provision are also required.

To overcome many of these problems parallel programming systems are often helpful. They make parallel programs construction easier by providing a higher

© The Author(s), under exclusive license to Springer Nature Switzerland AG 2023
V. Malyshkin (Ed.): PaCT 2023, LNCS 14098, pp. 42–50, 2023.
https://doi.org/10.1007/978-3-031-41673-6_4

level programming abstractions to the user than those of conventional parallel programming and automate the process of parallel programs construction while taking care of many low-level efficiency-related problems. Particularly, dynamic properties are often provided by run-time systems.

One of the main problems of run-time systems use is the overhead they imply. In many cases the overhead can be times greater than the effective work (i.e. the computational workload excluding run-time system overhead or other work, related to communications, synchronization, etc.). Reduction of run-time systems overhead is a relevant problem for run-time systems.

Trace playback is a powerful technique for run-time systems overhead reduction [1]. It is applicable for the cases where a series of numerical experiments has to be run. The essence of the technique is to perform an ordinary execution of a parallel program under run-time system control while recording every effective action the run-time system decides to take. Such a recording is called *trace*. The rest of the executions of the series are performed without the run-time system. Instead, a lightweight trace playback system (TPS) is used. The TPS just reads the trace and reproduces the actions taken by the run-time system in the first round.

Trace playback technique is not new. In different forms the idea was employed in different fields. In databases trace playback is used to optimize query handling [2] (MS SQL [3], Oracle RDBMS [4], PostgreSQL [5]). Similar approach is used in Just-In-Time (JIT) compilation [6], used in Java [7], JavaScript [8], .NET framework [9] etc. [10] In hardware microarchitectures where (emulated) machine code is translated in run-time by processor to another (native) machine code, the translated machine code is cached and reused [11] (Transmeta Crusoe [12], QEMU [13], Intel Pentium 4 [14]). In parallel programming the technique is implemented in LuNA [1] programming system.

Trace playback technique possesses a possibility to further improve efficiency of parallel program execution. It consists in transforming a trace before execution to make it fit better into particular hardware. The modification may include changing distribution of workload to computing nodes, reordering operations, etc. The main two reasons to transform the trace ('balance it') are as follows. Firstly, even if the work was ideally balanced with normal (run-time based) execution, the trace may provide full load of hardware resources, since the trace is being executed in different conditions (i.e. absence of run-time system overhead). Secondly, if a trace has to be played back on another number of computing nodes, then the work imbalance is likely to occur. In both cases the trace can be modified in order to improve efficiency of the playback.

In this paper an improvement to the trace playback technique is proposed, which consists in balancing the trace before playback. The improved technique is implemented on the basis of LuNA system for parallel programs construction automation. This work continues the work [1], where the trace playback technique implementation for LuNA was proposed.

The rest of the paper is organized as follows. The next section provides necessary background on LuNA system and the basic trace playback technique implementation in the system. Section 3 proposes the improvement of the technique. Section 4 presents the experimental results. The conclusion closes the paper.

2   LuNA Computational Model and Trace Playback Technique

The trace playback technique cannot be implemented for a random programming language or a system. The computational model of the language or the system matters. Let us illustrate this by considering an example of the system LuNA as an example of a computational model which allows implementation of the technique.

LuNA (Language for Numerical Algorihtms) is a language and a system for numerical parallel programs construction automation [15]. It is an academic project of the Institute of computational mathematics and mathematical geophysics SB RAS. The system is based on computational models [16] and follows the active knowledge [17] approach.

In LuNA a program is considered as a description of a recursively enumerable set of triplets of form $\langle in, mod, out \rangle$, where in and out are correspondingly input and output immutable aggregated variables called *data fragments* (DF) and mod is a no side-effects computational module (e.g. a conventional serial subroutine). The triplets are called computational fragments (CF). Each CF computes values of out DFs from values of in DFs. LuNA program execution is execution of all its CFs according to the dataflow model. Immutability of DFs and the absence of side effects of the modules allows the system to distribute and redistribute CFs and DFs to computing nodes of a multicomputer, reorder CFs execution (unless information dependencies are violated), automate communications and do many other routines in order to provide the program execution and improve its efficiency.

It should be noted that LuNA program is a finite sequence of a symbols in a final alphabeth which describes a potentially infinite number of triplets in a parametric form. Execution of LuNA program on particular input data defines the particular set of triplets executed, and the set may be different for different input data. For example if a set of triplets is defined with a loop-like descriptor, then particular number of iterations may vary depending on input data.

Decisions on *behavior* of the program (i.e. distribution of CFs and DFs, CFs execution order, etc.) are made and implemented by LuNA run-time system, which is the source of the overhead. But in the end all the run-time system is effectively does is executing particular modules with particular arguments on particular computing nodes. If the sequence of such executions is recorded for each computing node (and each working thread) then it can be reproduced from this recording with the same computational result as if the run-time system was performing the exectution. This is the basic idea of trace playback implementation in LuNA system. Immutability of DFs and absence of module side effects is essential here.

Trace is essentially a sequence of tuples $\langle n, ts, mod, id_1, id_2, ..., id_k \rangle$. Each tuple describes the execution of a CF, where n is the computing node where execution took place, ts is a timestamp when the CF was executed, mod is a module, related to the CF, id_i are the identifiers of CF's input and output DFs.

Trace playback technique has a number of advantages.

1. It practically eliminates the run-time system overhead because no run-time system is running with trace playback. The TPS only implies a minor overhead, as compared to normal run-time system. Note, that the run-time system overhead is caused by necessity for the run-time system to dynamically make decisions, and dynamic decision making often requires extra communications, memory and computations.
2. Many unnecessary actions can be taken dynamically by a run-time system due to the absence of global and postmortem information at run-time. For example, dynamic load balancing often causes multiple migrations of a CF between nodes until it reaches the node on which a CF is computed. Or a DF transfer over network can take multiple hops while reaching the destination (dynamic routing). In case of trace playback all intermediate and unnecessary actions are omitted.
3. Garbage collection with trace playback can be simplified, since all actual data consumptions are directly seen in the trace.
4. While trace is static, the behavior may often be as efficient as if dynamic properties were actually provided, since the behavior was produced by run-time system decision making, although in previous program execution. For example, if during the first execution round a workload imbalance occurred, and the run-time system has managed to balance the load, the other execution rounds will execute more efficiently, if the imbalance was the same as in the first round. This is often the case with series of numerical experiments with similar input data.
5. It is possible to add some lightweight dynamic properties support to TPS in order to improve performance, but without the necessity to run a heavy run-time system.

However, there are also a number of essential drawbacks of the technique, which significantly reduce the possibilities of its application in practice.

1. It is generally impossible to playback a trace on different input data, since the effective work may not provide a valid execution of the program. This is the case when different input data corresponds to a different set of triplets executed. Conditional branches or indirect addressing are examples of program constructs which may affect the set of triplets to execute. However, the fact that the trace is not valid for particular input data can be relatively easy detected during playback. Also, in many practical cases input data do not affect the set of triplets executed and thus the technique can be safely used.
2. Trace playback presumes no dynamic decision making, thus the decisions on program behavior may be inadequate for different input data. Although this will not lead to incorrect outputs computation, the efficiency may be poor. This is likely to occur for programs with dynamic load imbalance.

3. Also inefficiency of the behavior recorded in the trace may be caused by the fact that the behavior related decisions were made in the conditions of the presence of the run-time system, while the playback is performed in the absence of the run-time system. At least partial elimination of this drawback is the main purpose of the proposed trace balancing technique.

3   Trace Balancing Technique

The main idea behind the proposed trace balancing technique is to analyze the trace to disclose probable workload imbalance, and to redistribute some of CFs from overloaded computing nodes to underloaded ones.

The trace balancing technique can be divided into two main steps: generation of corrective distribution of CFs to computing nodes and the modification of the trace in accordance with the corrective distribution of CFs.

The generation of corrective distribution of CFs to computing nodes also can be divided into several steps: gathering information of the computational load of cluster nodes, identifying load imbalance with further reduction of the imbalance by transferring CFs between computing nodes. The computational load on the computing nodes comprises CFs execution time periods. In order to determine the load imbalance a quantization-based algorithm was used. Corrective distribution generation algorithm is described below.

1. Collect information about CFs' execution time periods.
2. Apply quantization - divide the program running time into equal intervals, the value of which is equal to the *quantization step* (a parameter of the algorithm).
3. In each quantum for each node calculate computational load and average quantum load.
4. If there is a load imbalance in the quantum, then transfer CFs from the most loaded node to the least loaded node.

This algorithm does not consider information dependencies between CFs.

The next step is to modify the trace according to the corrective distribution of the CFs. Below we will consider a trace not as a single sequence, but as a finite set of traces, each comprising all records, related for particular computing node. The term *modification* here refers to the process of relocating CFs between cluster nodes' traces. If CFs are relocated incorrectly, the modified trace may enter a deadlock state during trace playback. The deadlock can occur if modification violates information dependencies. Let us denote *correct trace* a trace that is replayed without deadlocks. Maintaining trace correctness during trace modification is mandatory.

To ensure the trace remains correct during modification it is necessary to uphold the order of execution for CFs. It is convenient to use timestamps of the start of execution of CFs since the beginning of the execution of a CF implies that all its information dependencies have been fulfilled, thereby ensuring the absence of deadlocks. The expected problem here is desynchronization of clocks on different computing nodes. The problem is well-studied in literature, so we will only highlight that the problem is worth solving by shifting timestamps in the traces rather than trying to synchronize clocks in run-time, since even the tiniest desynchronization may lead to deadlocks.

Let us also note that the algorithm parameter (quantization step) significantly affects the effectiveness of the algorithm (Sect. 4 confirms that statement experimentally). Obtaining a suitable value of the parameter is a problem to be solved. In the work the parameter values were obtained empirically. To reduce the parameter value search time a simple trace playback emulator was developed.

Trace Playback Emulator. By having knowledge of the playback time for each CF, it becomes possible to emulate the TPS. The process of emulating trace playback differs from actual trace playback in that, instead of reproducing CF execution, the execution time is adjusted by increasing a timer. Additionally, it is necessary to consider the delays in data transmission between cluster's nodes, which are specific to each respective cluster.

The utilization of a TPS emulator considerably expedites the parameter search process for the corrective distribution generation algorithm. This is due to the faster emulation compared to actual trace playback and the absence of necessity to use multicomputer for emulation (e.g. a PC can be used). Based on the predicted data from the emulator, the most successful modified traces can be selected.

4   Experimental Study

To study the playback performance of balanced traces, a program simulating a self-gravitating dust-cloud using the Particle-In-Cell (PIC) method [18] was selected as an exemplary complex real-world task specifically designed for execution on supercomputers. This program serves as a valuable example for studying the playback performance of balanced traces.

The tests were conducted on the Information and Computing Center cluster of Novosibirsk State University[1]. The testing was performed on traces generated using various parameters of the corrective distribution generation algorithm (see Table 1). Task parameters: mesh size 50^3, number of particles 10^8. The task executions were carried out on 4 computing nodes, with 4 cores available on each node.

Table 1 presents the testing results, with LuNA representing the original program written in LuNA language. LuNA-TP refers to the execution time of the LuNA program using the trace playback technique. Balanced LuNA-TP indicates the playback time of a balanced LuNA-TP trace, considering different quantization step values.

From Table 1 it can be seen that trace playback technique does reduces the execution time, while trace balance technique further improves the performance.

---

[1] https://nusc.nsu.ru/.

Table 1. Comparative execution time with different quantization step

Execution time (sec.)						
LuNA	LuNA-TP	Balanced LuNA-TP (quantization step, s.)				
		0.5 s	1 s	5 s	10 s	15 s.
601.185	136.055	127.94	115.685	110.503	100.808	136.340

Quantization step significantly affects the execution time with the optimal value of 10 s (among tested ones).

The following tests (Table 2) were conducted on a program where all CFs and DFs were distributed to a single computing node while other 3 nodes were idle. Trace balancing technique thus was used to generate fragments' distribution to computing nodes. This experiment emulates the case where the run-time system fails to construct a reasonable distribution of fragments to nodes. Trace balancing technique, however, still can be used to improve the distribution for trace playback. From the tests it can be seen that the best result was reached for quantization step of 5 s.

Table 2. Balancing a deliberately imbalanced fragments' distribution

Execution time (sec.)					
LuNA	LuNA-TP	Balanced LuNA-TP (quantization step, s.)			
		0.5 s	1 s	5 s	10 s.
509.284	184.139	204.139	197.857	157.402	158.247

The next tests were devoted to the usage of the trace balance technique to subsequently improve the efficiency. The idea behind the experiment is that trace playback itself can be used to produce a new trace, which can be balanced again, etc. Since all timings (CFs execution times, DFs transfer times, etc.) will normally change with trace playback as compared to run-time system execution, new imbalances can emerge (and be balanced).

In Table 3 the subsequent trace balances and playback times are shown. The 1st iteration is the same as in Table 2. The best time at quantization step of 5 s is selected for the next iteration. I.e. a new trace was recorded while executing the 1st iteration, and it was balanced with different quantization steps (the 2nd iteration row). The best result (at quantization step of 10 s) was selected for the third iteration. The execution time was reduced further to 136.402 s.

This test demonstrates that trace balancing technique can be used iteratively to improve trace playback efficiency.

Table 3. Trace execution time with iterative trace balancing

Quantization step	0.5 s.	1 s.	5 s.	10 s.
1st iteration	204.139	197.857	157.402	158.247
2nd iteration	202.446	182.182	151.841	145.109
3rd iteration	173.843	175.532	139.715	136.402

5   Conclusion

Trace playback technique is a promising approach to eliminating run-time system overhead, which can be implemented for suitable parallel programming systems, and employed in some computational experiments. In the paper an improvement over the technique is proposed, which is based on balancing traces before playback. The improved technique was implemented on the basis of LuNA system for automatic parallel programs construction. The tests showed that the proposed improvement is effective at least in some practical cases, especially when the run-time system fails to provide satisfactory workload distribution.

Further study of the topic may include development of validation methods to ensure the trace is valid for given input data. Trace balancing techniques should be improved to consider communications impact. Here many existing cost models and scheduling algorithms can be employed. Another promising study topic is generation of imperative parallel programs from a trace (e.g. a C++/MPI program). Also dynamic trace balancing algorithms can be employed with trace playback. The latter is a trade-off between dynamic properties provision and run-time system overhead.

Acknowledgements. The work was supported by the budget project of the ICMMG SB RAS No.0251-2022-0005 and partially funded by the Science Committee of the Ministry of Science and Higher Education of the Republic of Kazakhstan [Grant No. AP09058423].

References

1. Malyshkin, V., Perepelkin, V.: Trace-based optimization of fragmented programs execution in LuNA system. In: Malyshkin, V. (ed.) PaCT 2021. LNCS, vol. 12942, pp. 3–10. Springer, Cham (2021). https://doi.org/10.1007/978-3-030-86359-3_1
2. Viglas, S.D.: Just-in-time compilation for SQL query processing. In: IEEE 30th International Conference on Data Engineering. Chicago, IL, USA, pp. 1298–1301 (2014). https://doi.org/10.1109/ICDE.2014.6816765
3. SQL Server Distributed Replay. https://learn.microsoft.com/en-us/sql/tools/distributed-replay/sql-server-distributed-replay?view=sql-server-ver16. Accessed 01 May 2023
4. Galanis, L., et al.: Oracle database replay. In Proceedings of the 2008 ACM SIGMOD International conference on Management of data (SIGMOD '08). Association for Computing Machinery, New York (2008). https://doi.org/10.1145/1376616.1376732

5. Pgreplay – record and replay real-life database workloads. https://github.com/laurenz/pgreplay. Accessed 01 May 2023

6. Aycock, J.: A brief history of just-in-time. ACM Comput. Surv. **35**(2), 97–113 (2003). https://doi.org/10.1145/857076.857077

7. Krall, A.: Efficient JavaVM just-in-time compilation. In: Proceedings. 1998 International Conference on Parallel Architectures and Compilation Techniques (Cat. No.98EX192), Paris, France, pp. 205–212 (1998). https://doi.org/10.1109/PACT.1998.727250

8. Ha, J., Haghighat, M.R., Cong, S., McKinley, K.S.: A concurrent trace-based just-in-time compiler for single-threaded javascript. Proc. PESPMA (2009)

9. Bebenita, M., et al.: SPUR: a trace-based JIT compiler for CIL. In Proceedings of the ACM International Conference on Object Oriented Programming Systems Languages and Applications (OOPSLA '10), pp. 708–725. Association for Computing Machinery, New York (2010). https://doi.org/10.1145/1869459.1869517

10. Izawa, Y., Masuhara, H.: Amalgamating different JIT compilations in a meta-tracing JIT compiler framework. In Proceedings of the 16th ACM SIGPLAN International Symposium on Dynamic Languages (DLS 2020), pp. 1–15. Association for Computing Machinery, New York (2020). https://doi.org/10.1145/3426422.3426977

11. Böhm, I., von Koch, T., Kyle, S.C., et al.: Generalized just-in-time trace compilation using a parallel task farm in a dynamic binary translator. SIGPLAN Not. **46**(6), 74–85 (2011). https://doi.org/10.1145/1993316.1993508

12. Dehnert, J.: The transmeta code morphing software: using speculation, recovery, and adaptive retranslation to address real-life challenges. In: Dehnert, J., Grant, B., Banning, J., Johnson, R., et al. (eds.) Proceedings of the International Symposium on Code Generation and Optimization (2003)

13. QEMU: a generic and open source machine emulator and virtualizer. https://www.qemu.org/. Accessed 01 May 2023

14. Boggs, D., et al.: The microarchitecture of the intel pentium 4 processor on 90 nm technology. Intel Technol. J. **8**(1) (2004)

15. Malyshkin, V.E., Perepelkin, V.A.: LuNA fragmented programming system, main functions and peculiarities of run-time subsystem. In: Malyshkin, V. (ed.) PaCT 2011. LNCS, vol. 6873, pp. 53–61. Springer, Heidelberg (2011). https://doi.org/10.1007/978-3-642-23178-0_5

16. Valkovsky, V.A., Malyshkin, V.E.: Synthesis if parallel programs and system on the basis of computational models, Nauka, Novosibirsk, p. 128 (1988). In Russian: Valkovsky, V.A., Malyshkin, V.E. (eds.) Sintez parallelnykh program i system na vychislitelnykh modelyah. Nauka, Novosibirsk, 128 str (1988)

17. Malyshkin, V.: Active Knowledge, LuNA and literacy for oncoming centuries. LNCS **9465**, 292–303 (2015)

18. Kireev, S.: A parallel 3D code for simulation of self-gravitating gas-dust systems. In: Malyshkin, V. (ed.) PaCT 2009. LNCS, vol. 5698, pp. 406–413. Springer, Heidelberg (2009). https://doi.org/10.1007/978-3-642-03275-2_40

Case Study for Running Memory-Bound Kernels on RISC-V CPUs

Valentin Volokitin, Evgeny Kozinov, Valentina Kustikova, Alexey Liniov, and Iosif Meyerov[✉]

Lobachevsky State University of Nizhni Novgorod, 603950 Nizhni Novgorod, Russia
meerov@vmk.unn.ru

Abstract. The emergence of a new, open, and free instruction set architecture, RISC-V, has heralded a new era in microprocessor architectures. Starting with low-power, low-performance prototypes, the RISC-V community has a good chance of moving towards fully functional high-end microprocessors suitable for high-performance computing. Achieving progress in this direction requires comprehensive development of the software environment, namely operating systems, compilers, mathematical libraries, and approaches to performance analysis and optimization. In this paper, we analyze the performance of two available RISC-V devices when executing three memory-bound applications: a widely used STREAM benchmark, an in-place dense matrix transposition algorithm, and a Gaussian Blur algorithm. We show that, compared to x86 and ARM CPUs, RISC-V devices are still expected to be inferior in terms of computation time but are very good in resource utilization. We also demonstrate that well-developed memory optimization techniques for x86 CPUs improve the performance on RISC-V CPUs. Overall, the paper shows the potential of RISC-V as an alternative architecture for high-performance computing.

Keywords: High-Performance Computing · RISC-V · ISA · C++ · Performance Analysis and Optimization · Memory-Bound Applications

1 Introduction

The development of new CPU architectures significantly affects the progress of computer technologies and their application in computer-aided design and engineering. At first glance, there has already been a variety of CPU architectures that satisfy many current needs. These architectures have gone through a thorny path from the first ideas and experimental samples to full-fledged products mass-produced by leading microprocessor manufacturers. It is also necessary to take into account the difficulties with the deployment of new devices, the development of a full stack of specific software and growing of educational ecosystem. All these things are quite expensive and very difficult. However, attempts to freeze progress and settle for only incremental improvements will by no means lead to any significant breakthroughs that people need. Additionally, the closeness and commercial ownership of the existing proprietary architectures (x86,

© The Author(s), under exclusive license to Springer Nature Switzerland AG 2023
V. Malyshkin (Ed.): PaCT 2023, LNCS 14098, pp. 51–65, 2023.
https://doi.org/10.1007/978-3-031-41673-6_5

ARM, and others) leads to complicated problems and limitations. Technologies controlled by large companies are usually closed to change, which reduces the potential for further development.

The project of a new free and open architecture RISC-V [1, 2] based on the RISC (Reduced Instruction Set Computer) concept [3] which was presented more than 10 years ago at the University of California at Berkeley deserves attention. In just 12 years hardware and software developers have managed to introduce quite efficient CPUs, publicly available for purchase and use. The performance of existing RISC-V CPUs is still far even from mobile x86 and ARM CPUs, but progress in this area is proceeding at a significant pace. It is unlikely that anyone dares to predict when the first high-performance RISC-V CPU will be created, but the prospects look quite real, and it is confirmed by the current announcements of developers, the investments of industry leaders (for example, Intel), and the growing interest of the community [4].

In this paper, we analyze the performance of two available RISC-V devices in solving problems in which memory management is the main factor affecting computation time. Our main interest is to assess current opportunities and future prospects and answer the following key questions:

- What are the opportunities to adapt existing system software to work on RISC-V CPUs, and what efforts are needed?
- What performance indicators related to the memory subsystem of available RISC-V devices are achievable on standard benchmarks that are commonly used on x86 and ARM architectures?
- How does the attainable performance of RISC-V devices compare with the peak performance specified by the hardware manufacturers?
- Are well-established memory optimization techniques applicable to improve performance on RISC-V devices?

To get the first answers to these questions, we tested performance on two RISC-V devices (Mango Pi MQ-Pro and StarFive VisionFive), an ARM device (Raspberry Pi 4 model B), and a high-performance Intel Xeon 4310T server. We study performance on the following memory-bound benchmarks: a standard STREAM test [5], an in-place dense matrix transpose algorithm, and an image filtering algorithm. On the STREAM benchmark, we determined the memory bandwidth for each of the devices. Using the implementations of transposition and filtering algorithms, we studied how well-known techniques for optimizing performance by improving the reuse of data loaded into a cache affect the computation time on RISC-V devices. When analyzing the results, we paid not so much attention to comparing the total computation time, which obviously looks unfair, but to studying how efficient the utilization of available computing resources is. To the best of our knowledge, this work is at least one of the first papers analyzing the performance of algorithms on RISC-V devices.

2   RISC-V Architecture

RISC-V is an open instruction-set architecture (ISA) that was originally designed for research and education [1, 2, 6–8]. It is developed from scratch, taking into account the shortcomings of other open architectures and free from the issues of proprietary architectures that are forced to maintain backward compatibility. RISC-V avoids "over-architecting" for a particular microarchitecture style and has a modular design, a compact base ISA and many extensions [2, 8]. So it can be used in systems of any complexity up to high-performance devices like manycore CPUs or accelerators. Support for IEEE-754 floating-point standard, extension for vector operations, privileged and hypervisor architecture allow us to develop both conventional and HPC applications, operating systems, and virtualization software.

The main difference between RISC-V and other popular architectures, such as Intel or ARM, is primarily its openness and free availability. Additionally, it is distinguished by the small size of the basic instruction set (47 instructions only), modular extension principle, fixed base set and extensions after their standardization. All of this allows for relatively easy build of architectures for different application areas. The ability to run Linux on almost any RISC-V system and standardized requirements for running Android give hope that the architecture will be popular in mobile and IoT devices, infrastructure, industrial and HPC systems.

Architecture authors develop the corresponding CPU core microarchitectures, processors, and complete systems. Since 2011, when the "Raven 1" SoC was created (ST 28 nm FDSOI), they have released a number chips of Raven [9, 10], Hurricane [11, 12], Craft, Eagle, BROOM [13] families. The latter of these use the BOOM microarchitecture [14–16], which is structurally similar and performance competitive with commercial high-performance out-of-order cores. CPU and system-on-a-chip implementations of the architecture are performed by tens of companies and ranged from microcontrollers to high-performance cluster prototype [17]. The number of RISC-V processor cores shipped to date exceeds 10 billions [18].

The RISC-V software stack includes all the necessary tools for application development. The operating system (Linux), compiler (gcc, Clang), core libraries and programming tools are available for every existing implementation. Current prototypes support a limited set of HPC technologies, namely OpenMP, a set of base libraries (openmpi, openBLAS, fftw), and a set of applications which can be compiled and built on RISC-V (WRF, BLAST, GROMACS, VASP, and others) [19, 20]. However, the interest and support from the academic community, commercial companies and government organizations [20–23] will likely bring RISC-V systems to the level of high-performance solutions in the near future.

3 Benchmarking Methodology

3.1 Infrastructure

We employed two currently available RISC-V devices:

1. Mango Pi MQ-Pro (D1) with Allwinner D1 processor (1 x XuanTie C906, 1 GHz) and 1 GB DDR3L RAM. Ubuntu 22.10 operating system (RISC-V edition) and GCC 12.2 compiler were installed.

   Some architectural features of C906 are as follows: RV64IMAFDCV ISA, 5-stage single-issue in-order execution pipeline, L1 2-way set-associative I-Cache and 4-way set-associative D-Cache with a size of 32 KB each and cache line size of 64 bytes, Sv39 MMU, fully associative L1 uTLB with 20 entries (10 I-uTLB and 10 D-uTLB), 2-way set-associative L2 jTLB of 128 entries, gshare branch predictor with 16 KB BHT, hardware prefetch for instructions (the next consecutive cache line is prefetched) and data (two prefetch methods: forward and backward consecutive and stride-based prefetch with stride less or equal 16 cache lines), 16, 32, 64-bits integer and fp scalar and 512-bit vector operation including fp FMA.

2. StarFive VisionFive (v1) with StarFive JH7100 processor (2 x StarFive U74, 1 GHz) and 8 GB LPDDR4 RAM. OS Ubuntu 22.10 (RISC-V edition) and GCC 12.2 compiler were installed.

Some architectural features of U74 core are as follows: RV64IMAFDCB ISA, 8-stage dual-issue in-order execution pipeline, L1 2-way set-associative I-Cache and 4-way set-associative D-Cache with a size of 32 KB each, cache line size of 64 bytes and a random re-placement policy (RRP), 128 KB 8-way L2 cache with a RRP, hardware data prefetch (forward and backward stride-based prefetch with large strides and automatically increased prefetch distance), Bare and Sv39 MMU modes, fully associative L1 ITLB and DTLB with 40 entries each, direct mapped L2 TLB with 512 entries, branch predictive hardware with 16-entry BTB and 3.6 KB BHT, 64-bits integer and 32, 64-bits floating point scalar operation including fp FMA.

To compare results, we also used a Raspberry Pi device (ARM) and an Intel Xeon server processor (x86) with the following configuration:

1. Raspberry Pi 4 model B with Broadcom BCM2711 (4 x Cortex-A72, up to 1.5 GHz) processor and 4 GB LPDDR4 RAM. Ubuntu 20.04 operating system and GCC 9.4 compiler were installed.

2. Server with 2 x Intel Xeon 4310T (2 × 10 Ice Lake cores, up to 3.4 GHz) and 64 GB DDR4 RAM. CentOS 7 operating system and GCC 9.5 compiler were installed. Only the cores of the first CPU were used to eliminate the occurrence of NUMA effects that are obviously absent on other devices.

A direct comparison of powerful server hardware and low-power devices may seem patently disadvantageous for the latter, however, we decided to include the x86 server in the comparison because performance results on a x86 server look reasonable and expected by the HPC community.

3.2 Benchmarks

In this paper, we study the performance of RISC-V devices on three memory-bound benchmarks. First, we experimentally determine the memory bandwidth using the commonly applied STREAM benchmark [1], which performs elementary operations on vectors. Measuring memory bandwidth allows us to interpret the results of subsequent experiments on x86, ARM and RISC-V devices and compare them, taking into account the capabilities of the hardware. Next, we consider the in-place matrix transpose algorithm, which is one of the basic dense linear algebra algorithms. We present several implementations of the algorithm and check how typical memory optimization techniques performing well on x86 and ARM CPUs work on RISC-V devices. Finally, we follow the same idea in studying the Gaussian Blur algorithm, successively applying different approaches to code optimization and testing to what extent they speed up calculations on different devices, with a particular focus on RISC-V devices. Summarizing the obtained results, we formulate the main conclusions about the future prospects for using RISC-V devices in HPC applications.

3.3 Performance Metrics

The computation time is typically used as the main performance metric. However, we should keep in mind that our comparison involves a single-core low-power processor of the RISC-V architecture, which is still at the beginning of its development, and a 10-core powerful Xeon processor that uses many advances in the field of high-performance computing. Therefore, in addition to the computation time, we also used relative metrics that allow us to make a fair comparison in terms of the utilization of available computational resources.

Given that the RISC-V architecture is still experimental, it is not clear which of the optimization techniques typical for x86 and ARM work well on RISC-V. Therefore, the following question is of interest: what kind of improvement from a naïve version of the algorithm can be obtained by performing a series of memory optimizations typical for conventional CPUs. This metric allows us to understand what kind of improvement can be leaded to by a certain optimization in each particular case, supplementing the computation time, which is dependent from a device features.

Another metric we employ allows us to evaluate how efficiently we use the available memory channels. To evaluate this, we introduce the following metric. At first we calculate the ratio of the data number of bytes that needs to be moved between DRAM and CPU to the computation time. We divide the value calculated in this way by the achieved memory bandwidth, measured by the STREAM benchmark. The result belongs to the segment from zero to one and is dimensionless. The closeness of this value to one indicates that the algorithm uses the bandwidth of the memory channels quite rationally. This metric allows us to compare the devices, taking into account their significantly different performance of the memory subsystem.

Overall, absolute (computation time, memory bandwidth) and relative (speedup from a naive implementation, "utilization of a memory subsystem") metrics allow us to draw conclusions about the current state of the considered RISC-V devices.

4 Numerical Results

4.1 STREAM Benchmark

The STREAM benchmark [1] is one of the popular ways to measure achieved memory bandwidth. Like other similar benchmarks, STREAM is based on the idea of reading and writing an array of data from the memory of the corresponding level. STREAM uses 4 tests that have different bytes/iter and FLOPS/iter values:

1. COPY – is simple copying from one array to another (a[i] = b[i]). This operation transfers 16 bytes per iteration and does not perform floating point calculations.

2. SCALE – is copying from one array to another with multiplication by a constant (a[i] = d * b[i]). This operation transfers 16 bytes per iteration and does 1 FLOPS/iter.
3. SUM – the sum of elements from two arrays is stored in the third array (a[i] = b[i] + c[i]). This operation transfers 24 bytes per iteration, but still does 1 FLOPS/iter.
4. TRIAD – FMA (fused multiply-add) from elements from two arrays (a[i] = b[i] + d * c[i]) is stored in the third array. This operation transfers 24 bytes per iteration and does 2 FLOPS/iter.

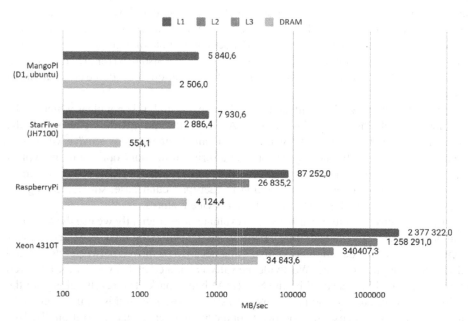

Fig. 1. The results of the STREAM benchmark

We select the sizes of the arrays in such a way that they are not forced out of the memory of the considered level and could not be cached efficiently in faster memory. All levels of memory that are available on each specific device are considered. We run a multi-threaded (for a shared memory) or sequential (for an individual resource, for example, an L1 cache) version of STREAM. In the sequential experiments the results are multiplied by the number of cores. Overall, we use the maximum value that is achieved during sufficiently large number of repetitions of the experiment.

The results of the obtained throughputs are presented in Fig. 1. It turned out that the RISC-V devices have a number of drawbacks. We found that there is only L1 cache with a rather low bandwidth on the Mango Pi board with the Allwinner D1 processor compared to other devices. In the case of the StarFive board on JH7100, we observe the low bandwidth of DRAM which corresponds to the reduced memory channel in the device. Overall, the memory subsystem of the RISC-V devices is behind its analogue on ARM and, as expected, is even more inferior to the x86 Xeon CPU.

4.2 In-Place Dense Matrix Transposition Algorithm

Algorithm. The in-place dense matrix transposition algorithm is one of the key algorithms in linear algebra [24]. It is used both as a standalone procedure and as part of other linear algebra algorithms. In this section, we consider a sequence of optimizations to the matrix transposition algorithm that incrementally improve performance from the most basic (naïve) implementation to the efficient high-performance version.

Naïve Implementation ("Naïve"). The following implementation (Listing 1) is a code that a programmer often develop without thinking about performance optimization. Of course, such an implementation cannot be expected to perform efficiently, but it is inefficient for all the devices. All are on equal terms.

```
1:   Transpose_baseline (double * mat, int size)
2:      for (i=0; i < size; i++)
3:         for (j=i+1; j < size; j++)
4:            mat[i][j] = mat[j][i]
```

Listing 1. Pseudocode of the naïve implementation

Parallelization ("Parallel"). Since most modern hardware is multi-core, using multithreading is an important way to reduce the computation time. In this algorithm, the iterations of the outer loop are independent of each other. Therefore, the algorithm can be easily parallelized using the OpenMP technology. Note that OpenMP is supported by all compilers on 4 considered devices. However, the Allwinner D1 (Mango Pi) device is single-core, so it makes no sense to use parallelism there, and other optimizations for this device are performed in sequential code.

Better Data Reuse: Cache Blocking ("Blocking"). The next optimization is to avoid unnecessary data loads and better reuse of data already loaded into caches. This can be achieved by block traversal of a matrix, which is typical for many matrix algorithms. Listing 2 shows the pseudocode without unnecessary implementation details.

```
1:   Transpose_block (double * mat, int size)
2:     parallel_for (i_blk=0; i_blk<size; i_blk+=blk_size)
3:       for (j_blk=i_blk; j_blk<size; j_blk+=blk_size)
4:         for (i=i_blk; i<i_blk+blk_size; i++)
5:           for (j=j_blk +1; j<j_blk+blk_size; j++)
6:             mat[i][j] = mat[j][i]
```

Listing 2. Pseudocode of the block algorithm implementation

Improved Memory Access ("Manual_blocking"). The next optimization continues and enhances the ideas of the previous one. Its main task is to provide, if it is possible, sequential access to RAM. To do this, blocks are loaded into the cache manually, after which they are transposed and data is exchanged with other blocks. The pseudocode is shown in Listing 3.

```
1:   Transpose_improvedMemAccess (double * mat, int size)
2:     parallel_for(i_blk=0; i_blk<size; i_blk+=blk_ size)
3:       double cache_blk[blk_size*blk_size]
4:       for (j_blk=i_blk; j_blk<size; j_blk+=blk_size)
5:         load_block_to_cache (i_blk, j_blk)
6:         transpose_block_in_cache()
7:         swap_block (j_blk, i_blk)
8:         transpose_block_in_cache()
9:         store_block (i_blk, j_blk)
```

Listing 3. Pseudocode of the improved block algorithm implementation

Dynamic Scheduling ("Dynamic"). The final version of the code differs from the previous one in the dynamic scheduling of the parallel loop. It makes it possible to eliminate the imbalance in the computational load that occurs in traversing the rows of an upper triangular matrix, which obviously have different lengths.

Performance Results and Discussion. Figure 2 shows the computation time of the presented algorithms on four devices. In accordance with the previously introduced metrics, it shows the computation time of the naïve version of the algorithm on different devices, as well as the acceleration of optimized implementations relative to the naive version for each of the platforms. The lack of acceleration of parallel implementations (Parallel and Dynamic versions) on Mango Pi is due to the single-core CPU.

We found that optimizations that were developed for the x86 architecture perform well also on RISC-V devices. Despite significant architectural differences between the devices, the memory subsystems are organized with the similar principles, so optimizations have made it possible to better utilize memory resources of RISC-V, ARM and x86 CPUs. Note that the presented optimizations show a good acceleration, especially considering that this algorithm does not use vector instructions, which in many cases can speed up calculations and make working with memory more efficient.

Given the substantially larger computing capabilities of Intel Xeon, we compare the overall computation time on RISC-V and Raspberry Pi CPUs. Note that despite the very large advantage of the latter in memory bandwidth at the STREAM benchmark over both RISC-V devices, the gap in computation time between RISC-V and ARM is much smaller. Moreover, with an increase of the matrix size to 16384, the difference in speedup compared to the naïve version on ARM and RISC-V CPUs decreases. It confirms better utilization of available resources of RISC-V CPU.

Comparing the results of two RISC-V boards with each other, we noticed that despite the good memory bandwidth on the device with the Allwinner D1 (Mango Pi) processor compared to the second one on the JH7100, their computation time is almost identical. To find out the reason for this phenomenon and analyze the performance in terms of one of the previously announced metrics, we calculated the relative memory bandwidth utilization (Fig. 3). This metric shows how efficiently the reuse of data loaded from memory is implemented, and how significantly the computation time depends on the properties of the memory subsystem. The optimal value of this metric, equal to one, is not achievable in many cases, but closeness to one indicates efficient memory utilization.

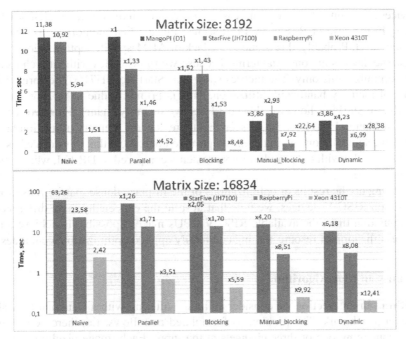

Fig. 2. Computation time of five implementations of the matrix transposition algorithm on four computing devices (Intel Xeon server, Raspberry Pi and two RISC-V boards). The labels above the bars of the diagram show the computation time of the naïve version of the algorithm given in seconds, as well as the speedup of the optimized implementations relative to the naïve ones on the corresponding devices. The top panel contains the results for a 8192 x 8192 matrix, the bottom one for a 16384 x 16384 matrix. The bottom panel does not contain results for Mango Pi because the matrix does not fit in memory of this device.

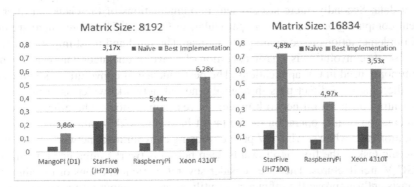

Fig. 3. Effectiveness of the relative memory bandwidth utilization for four devices. The metric is calculated for the naïve implementation and the best optimized implementation for each of the devices. The matrix size is 8192 x 8192 (left panel) and 16384 x 16384 (right panel). The right panel does not show results for the Mango Pi because the matrix does not fit in memory.

All devices show almost the same increase in this relative indicator for sufficiently large matrices. In the case of Raspberry Pi, it seems unusual that memory utilization is at such a low level. Probably, this is due to the lack of ARM-specific optimizations, but in this case the devices are on equal terms, because we run C codes without architecture-specific optimizations only. The metrics show that StarFive (JH7100) performed well in terms of memory bandwidth utilization. This is primarily due to the low memory bandwidth on StarFive, however, this board has two memory channels for two cores. In the case of Mango Pi (D1), it can be seen that there is a low memory utilization both in the naïve implementation and in the most optimized one. Note that this device has only one level of cache with only modest improvements compared to DRAM, which affects the performance.

Summing up the results of optimizations of the in-place dense matrix transposition algorithm on RISC-V devices, we note that despite the expectedly large difference in the computation time, the available RISC-V CPUs make it possible to achieve a high degree of utilization of resources using commonly applied optimization techniques.

4.3 Gaussian Blur Algorithm

Algorithm. In this section we consider an image filtering with a Gaussian Blur algorithm as a benchmark. The problem is formulated as follows. Let there be an image (tensor) containing one or three channels at the input. Each image pixel contains one or three intensity values, respectively, each in the range from 0 to 255, or from 0 to 1, if normalization is performed. The problem of filtering involves passing through the image from left to right and from top to bottom, applying the Gaussian filter kernel to the pixels, and calculating a discrete convolution. The output is an image that has the same spatial dimensions as the input one and contains updated intensity values.

We chose the filtering task as a benchmark for the following reasons. Firstly, it is necessary in many computer vision (CV) algorithms for the preliminary preparation of input data. Secondly, there are efficient implementations of the Gaussian filter for different computing architectures, in particular, in OpenCV. Therefore, there are optimized implementations to compare performance. The third and most important reason is that discrete convolution is a basic operation of convolutional neural networks, which are commonly used in CV applications. The performance of convolutions significantly affects the time of a direct pass through the neural network, which is critical in the implementing deep neural network models in real applications. The efficiency of the implementation of this operation on the target hardware highly influences the overall computation time when solving CV problems using convolutional networks. Therefore, the filtering task is the first step towards deep neural networks inference optimization on RISC-V architectures. Then we consider several implementations of the algorithm that consistently improve the efficiency of utilization of computing resources, and, as before, analyze the achieved performance results.

Naïve Implementation ("Naïve"). As a basic implementation, we use an algorithm in which the Gaussian filter kernel is used to sequentially calculate the intensities of each pixel of the resulting image row by row (Listing 4).

```
1 :  cntChannel = 3
2 :  middle = sizeFilter / 2
3 :  for (i = 0; i < h - sizeFilter; i++)
4 :    for (j = 0; j < w - sizeFilter; j++)
5 :      for (c = 0; c < cntChannel; c++)
6 :        sum = 0.f
7 :        for (i_f = 0; i_f < sizeFilter; i_f++)
8 :          for (j_f = 0; j_f < sizeFilter; j_f++)
9 :            pos_i = (i + i_f) * (w * cntChannel)
10:            pos_j = (j + j_f) * (cntChannel) + c
11:            sum += srcData[pos_i + pos_j] *
                      filter[i_f * sizeFilter + j_f]
12:        i_d = i + middle; j_d = j + middle
13:        distData[(i_d*w+j_d)*cntChannel + c] = sum
```

Listing 4. Pseudocode of the naïve implementation of the Gaussian Blur algorithm

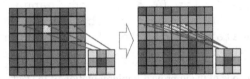

Fig. 4. Gaussian Blur filter optimization for color images: unit-stride memory access. Left panel: naïve implementation. Right panel: improved implementation.

Unit-stride Access ("Unit-stride"). Note that if a color image is used, then memory access is not unit-stride (Fig. 4, left panel). As a first modification, we change the order of the loops so that the loop through the image channels (line 5) is inside the filter kernel application loop (line 8). As a result, memory access is unit-stride (Fig. 4, right panel), which is much better in terms of memory usage.

One-dimensional Kernels ("1D_kernels"). For further optimization, we rearrange the computations [25] based on the following representation of the Gaussian filter:

$$G(x, y) = \frac{1}{2\pi\sigma^2}e^{-\frac{x^2+y^2}{2\sigma^2}} = \left(\frac{1}{\sqrt{2\pi}\sigma}e^{-\frac{x^2}{2\sigma^2}}\right)\left(\frac{1}{\sqrt{2\pi}\sigma}e^{-\frac{y^2}{2\sigma^2}}\right) \tag{1}$$

Now we can successively apply two one-dimensional Gaussian filter kernels instead of using a two-dimensional kernel (see Fig. 5).

The use of one-dimensional filters reduces the computational complexity of the entire algorithm. When using two-dimensional filters, the complexity can be estimated as $O(W \cdot H \cdot C \cdot F^2)$, where W and H are the width and height of the image, respectively, C is the number of channels, and F is the size of the filter kernel. By using two one-dimensional filters, the complexity can be reduced to $O(W \cdot H \cdot C \cdot F)$.

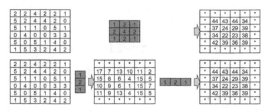

Fig. 5. Applying of the two-dimensional kernel (top row) and two one-dimensional kernels (bottom row) of the Gaussian blur filter

Improving Memory Access ("Memory"). In the previous implementation, the performance of applying the horizontal kernel of the Gaussian filter is low due to an inefficient memory access pattern. Therefore, we use the order of loops, in which each element of the kernel interacts with the entire row from the original image matrix (Listing 5).

```
1:   for (i = 0; i < h - sizeFilter; i++)
2:     for (i_f = 0; i_f < sizeFilter; i_f++)
3:       pos_i = (i + i_f) * (w * cntChannel)
4:       for (j = 0; j < w * cntChannel; j++)
5:         tmpData[(i + middle) * w * cntChannel + j] +=
             srcData[pos_i + j] * filter1D[i_f]
```

Listing 5. Pseudocode of the improved implementation of the Gaussian Blur algorithm

Parallel Implementation ("Parellel"). The computations are independent and well-balanced, therefore we parallelize the algorithm trivially by using #pragma parallel for from OpenMP.

Performance Results and Discussion. We used a color image of size 2544 × 2027 for the experiments. To apply the filter, the intensities of each pixel were converted to the float type. The size of the kernel of the Gaussian filter $F = 19$. The computation time and the speedups achieved are shown in Fig. 6. As before, the sizes of the bars correspond to the computation time of a particular algorithm on the corresponding device, while the captions above the bars show the computation time for the naïve version and speedup of other versions relative to the naïve implementation. The experimental results allow us to draw the following conclusions. Firstly, we found that the baseline implementation lags behind OpenCV[1] by several orders of magnitude, regardless of the device architecture. Then, the computation time of the first modification of the algorithm ("Unit-stride") is obviously better because of sequential memory access which us much faster due to an efficient data prefetch. Apparently, this is not the case for the StarFive device, where data prefetch does not speed up calculations because low memory bandwidth does not allow data to be prepared on time and only leads to additional overhead.

---

[1] In the case of using processors with RISC-V architecture, the OpenCV computation time was measured on a Linux image that supports vector instructions.

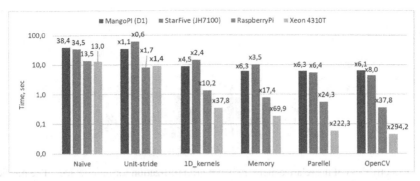

Fig. 6. Computation time of five implementations of the Gaussian Blur algorithm on four computing devices (Intel Xeon server, Raspberry Pi and two RISC-V boards). The labels above the bars of the diagram show the computation time of the naïve version of the algorithm given in seconds, as well as the speedup of the optimized implementations relative to the naïve ones on the corresponding devices.

Next, we paid attention to the computation time and speedup of the next modification of the algorithm ("1D_kernels"). As expected, the calculations are faster. It is worth noting in particular that with a filter size of $F = 19$, one would expect a substantial speedup. Apparently, it did not happen due to an excessive amount of memory access. This assumption is confirmed by the results of the following modification of the algorithm ("Memory"). Due to much more efficient memory access, the speedup compared to the naïve implementation becomes much larger. The acceleration by more than 19 times on the server with Intel Xeon 4310T processors is justified by the fact that the compiler has been able to vectorize the code with the loop order, used in the "Memory" implementation. The computation time and speedup of parallel implementations are shown in Fig. 6 ("Parallel" bar). Since the problem is memory bound, speedup is limited by the number of available memory channels.

Like for the matrix transposition benchmark, we analyze the effectiveness of memory bandwidth usage (see Fig. 7). When calculating this relative metric (see Sect. 3.3), we used a "1D_kernels" implementation as a baseline. We conclude that the memory subsystem of Mango Pi does not allow for high performance of the image filtering algorithm due to the lack of L2 cache and slow L1 cache. StarFive lags behind RaspberryPi in memory access performance, but overall, the results are comparable. In case of Intel Xeon 4310T, the parallel algorithm provided an increase in the memory bandwidth usage metric due to the presence of a larger number of memory channels, which are not available in other devices under consideration.

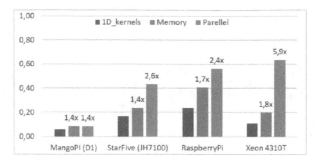

Fig. 7. Effectiveness of the relative memory bandwidth utilization for four devices. The metric is calculated for the three optimized implementations of the Gaussian Blur algorithm. Labels show the improvement compared to the "1D_kernels" implementation.

5 Conclusion

In this paper, we explored new opportunities and perspectives of the RISC-V computing architecture. Despite the many papers studying of architectural ideas and their possible implementations, testing existing RISC-V devices and studying their performance issues are of great interest. In this regard, the paper presents an analysis of the performance of two RISC-V devices on three memory-bound benchmarks in comparison with well-studied ARM and x86 CPUs. First, we measured the memory bandwidth using the commonly used STREAM benchmark. The results showed that the existing RISC-V prototypes are still significantly behind ARM and x86 devices. Therefore, when considering two benchmarks from linear algebra and image processing, we proposed to take into account not only the computation time, but also the efficiency of using the memory subsystem. It turned out that in the matrix transposition algorithm, RISC-V devices demonstrate excellent utilization of available resources, while when filtering images, memory is used less efficiently, which is caused by a small number of memory channels. We especially want to note that the typical memory optimization techniques worked out over the past decades on ARM and x86 behave as expected on RISC-V, allowing to significantly speed up calculations. As a result, we can conclude that although the available RISC-V devices are not yet suitable for HPC, but nevertheless they show a high potential for further development.

Acknowledgements. The study is supported by the Lobachevsky University academic leadership program "Priority-2030". Experiments were performed on the Lobachevsky supercomputer.

References

1. Asanović, K., Patterson, D.A.: Instruction sets should be free: the case for RISC-V. EECS Department. University of California, Berkeley. UCB/EECS-2014–146 (2014)
2. Waterman, A., Asanović, K.: The RISC-V instruction set manual, volume I: user-level ISA, document version 20190608-base-ratified. In: RISC-V Foundation (2019)
3. Furber, S.B.: VLSI RISC Architecture and Organization, 1st edn. CRC Press (1989)

4. Asanović, K.: Advancing HPC with RISC-V. In: Invited Talk at Supercomputing Conference (2022)

5. McCalpin, J.: Memory bandwidth and machine balance in current high performance computers. IEEE Comput. TCCA Newsl., 19–24 (1995)

6. History – RISC-V International. https://riscv.org/about/history/. Accessed 09 May 2023

7. Chen, T., Patterson, D.A.: RISC-V genealogy. EECS Department, University of California, Berkeley, Technical report UCB/EECS-2016-6 (2016)

8. Waterman, A., Asanovic, K., Hauser, J.: The RISC-V instruction set manual, Volume II: privileged architecture, document version 20211203 (2021)

9. Lee, Y., et al.: Raven: A 28nm RISC-V vector processor with integrated switched-capacitor DC-DC converters and adaptive clocking. In: 2015 IEEE Hot Chips 27 Symposium (HCS), pp. 1–45. IEEE (2015)

10. Zimmer, B., et al.: A RISC-V vector processor with tightly-integrated switched-capacitor DC-DC converters in 28nm FDSOI. In: 2015 VLSI Circuits, pp. 316–317. IEEE (2015)

11. Schmidt, C., et al.: Programmable fine-grained power management and system analysis of RISC-V vector processors in 28-nm FD-SOI. IEEE Solid State Circuits Lett. **3**, 210–213 (2020)

12. Wright, J.C., et al.: A dual-core RISC-V vector processor with on-chip fine-grain power management in 28-nm FD-SOI. IEEE Trans. VLSI Syst. **28**(12), 2721–2725 (2020)

13. Celio, C., et al.: BROOM: an open-source out-of-order processor with resilient low-voltage operation in 28-nm CMOS. IEEE Micro **39**(2), 52–60 (2019)

14. Zhao, J., et al.: SonicBOOM: the 3rd generation berkeley out-of-order machine. In: Fourth Workshop on Computer Architecture Research with RISC-V, vol. 5, pp. 1–7 (2020)

15. RISC-V BOOM. https://boom-core.org/. Accessed 09 May 2023

16. BOOM: The Berkeley out-of-order RISC-V Processor. https://github.com/riscv-boom. Accessed 09 May 2023

17. Bartolini, A., et al.: Monte cimone: paving the road for the first generation of RISC-V high-performance computers. In: 2022 IEEE 35th International System-on-Chip Conference (SOCC), pp. 1–6. IEEE (2022)

18. Europe steps up as RISC-V ships 10bn cores. https://www.eenewseurope.com/en/europe-steps-up-as-risc-v-ships-10bn-cores/. Accessed 09 May 2023

19. RISC-V Software Ecosystem Status. https://sites.google.com/riscv.org/software-ecosystem-status. Accessed 09 May 2023

20. Davis, J.D.: RISC-V in Europe: the road to an open source HPC stack. https://www.european-processor-initiative.eu/wp-content/uploads/2022/03/EPI-@-HPC-User-Forum.pdf. Accessed 09 May 2023

21. First International workshop on RISC-V for HPC. https://riscv.epcc.ed.ac.uk/community/isc23-workshop/. Accessed 09 May 2023

22. RISC-V ISA – MIPS. https://www.mips.com/products/risc-v/. Accessed 09 May 2023

23. Framework Partnership Agreement for developing a large-scale European initiative for HPC ecosystem based on RISC-V. https://eurohpc-ju.europa.eu/framework-partnership-agreement-fpa-developing-large-scale-european-initiative-high-performance_en. Accessed 09 May 2023

24. Chatterjee, S., Sen, S.: Cache-efficient matrix transposition. In: IEEE Proceedings Sixth International Symposium on High-Performance Computer Architecture, HPCA-6 (Cat. No. PR00550), pp. 195–205 (2000)

25. Moradifar, M., Shahbahrami, A.: Performance improvement of Gaussian filter using SIMD technology. In: International Conference on Machine Vision and Image Processing, pp. 1–6 (2020)

Frameworks and Services

Pair of Genes: Technical Validation of Distributed Causal Role Attribution to Gene Network Expansion

Diana Dolgaleva[2](✉)[iD], Camilla Pelagalli[4], Stefania Pilati[5][iD],
Enrico Blanzieri[3,4][iD], Valter Cavecchia[3], Sergey Astafiev[1][iD],
and Alexander Rumyantsev[1][iD]

[1] Institute of Applied Mathematical Research, Karelian Research Centre of RAS,
Pushkinskaya Str. 11, Petrozavodsk 185910, Russia
seryymail@mail.ru, ar0@krc.karelia.ru
[2] Petrozavodsk State University, Lenina Pr. 33, Petrozavodsk 185910, Russia
abcdi_do@mail.ru
[3] CNR-IMEM, Trento, Italy
{enrico.blanzieri,valter.cavecchia}@unitn.it
[4] DISI, University of Trento, Trento, Italy
camilla.pelagalli@studenti.unitn.it
[5] Research and Innovation Centre, Fondazione Edmund Mach,
San Michele all'Adige, TN, Italy
stefania.pilati@fmach.it

Abstract. The paper is dedicated to preliminary results and validation of the distributed solution for causal role attribution in gene network expansion problem. The key ingredients of the solution are the web application based on Shiny framework, RBOINC backend as an interface between R language and BOINC desktop grid framework, and parallel (multicore) implementation of the Peter-Clark (PC) algorithm for causal role attribution. The approach is technically validated on a gene network expansion problem for the grapevine (*Vitis vinifera*).

Keywords: Causal Relationship Discovery · Distributed Computing · RBOINC · Gene Network Expansion

1 Introduction

The computing experiments are an important part of bioinformatics, which is a well established field in the intersection of mathematics, computer science and biology. Among the tasks with high computational complexity are the problems in the field of genetics (e.g. reconstruction of the DNA from sequencing data), molecular biology (establishing the structure of a given protein, as well as drug discovery for a specific disease) and many others. In those fields, high computational power is needed, ranging from top supercomputers to the desktop grids, depending on the application field.

V. Malyshkin (Ed.): PaCT 2023, LNCS 14098, pp. 69–82, 2023.
https://doi.org/10.1007/978-3-031-41673-6_6

The genes are coded in the DNA and they are transcribed into mRNA that eventually is translated into proteins. It is nowadays possible to sequence the mRNAs and, by estimating their concentration, obtain data about the transcription level of each gene, namely their expression. The possible casual relationships between the expression levels of genes are important to determine and are usually studied by intervening in the system, however purely observational data (without intervention) are widely available, which makes it possible to perform *causal relationship discovery* in-silico. In particular, the study of gene expression regulators, such as transcription factors, is essential to reconstruct gene regulation processes and how they are altered under different experimental conditions. This information is represented by means of *gene regulatory networks*, namely graphs whose nodes are genes and edges indicate the presence of a chain of events that connect the expression level of one gene to another one.

Causal relation discovery from observational data to compute gene regulatory networks is a recent research topic that has been actively studied both theoretically and technically. In our case, observational data are represented by publicly available genome-wide gene expression compendia, that are collections of homogeneous normalized transcriptomic datasets. For the case study reported below, the dataset of a *Vitis vinifera* - the common grapevine - compendium has been used [25], but the same analysis is undergoing also for *Homo sapiens* and *Mus musculus* (mouse) transcriptomic datasets. Applying a causal inference algorithm, such as the PC algorithm, to this kind of data should be effective in catching causal-effect relationships among genes, thus providing hypotheses of molecular mechanisms and interactions occurring during development, disease, or any other biological process of interest. The in silico analyses need to be experimentally validated, but can contribute to accelerate novel discovery. The output is a directed graph where genes are the nodes and their interactions are represented as oriented edges. The overall idea of this approach is to refine a "first-level" co-expression network by applying a further step based on conditional independence to remove indirect or distantly related connections, thus producing the so-called association network. Then, and this is the focus of the work presented here, a third step computes the orientation of the edges, finally producing a gene regulatory network.

Due to the high computational complexity of the PC algorithm, this process is divided into two steps, both developed to be used in the more general framework of parallel and distributed computing. The first step, called OneGenE (One Gene Expansion), considers the whole transcriptomic datasets and applies a partial version of the PC algorithm to produce lists of associated genes without the direction information, called OneGenE expansion lists, for each gene of the dataset (Ref. [8,30]). This huge computation has been made possible by the implementation of a BOINC-based desktop grid. Then, here we present the second step, consisting in analyzing each expansion list by applying the complete version of the PC algorithm to the most associated genes to detect the causal-effect relationships among them thus producing oriented networks. To solve the problem of high computational need, we implemented a cloud web application with a BOINC desktop grid backend, and is validated using the grapevine data. Thus, the key contribution is the novel technical approach to causal role attribution which uses both parallel (multicore) and distributed (BOINC-based) computing to solve the computationally intensive task in the field of bioinformatics. A technical validation of the results is then presented, based on literature. Note that the approach is general purpose, and can be adopted to transcriptomic data of various origin, however, in the present paper we focus on the case study based on the grapevine dataset.

The structure of the paper is as follows. In Sect. 1.1 a brief literature survey is presented. A cloud application for causal attribution of gene regulatory network is discussed in Sect. 2, where, in particular, the details on the data acquisition and usage are given in Sect. 2.4. The results of technical validation of the approach are given in Sect. 3. The paper is finalized with a conclusion and discussion of some future research directions.

1.1   Related Literature

Here we briefly recall the literature related to causal relationship discovery and causal attribution. The key monograph sources for the causal relationship discovery are [13, 18, 23, 27, 29, 35], to name a few, with a focus on correlation in [18], nice theoretical background in [35], application focus to econometrics and epidemiology in [27], implementation focus of the graph-based and association-rules-based methods in [23], machine learning techniques in [29] and state-of-art in [13].

Note that many fields of bioinformatics and, in particular, the causal relationship discovery problem, are computationally intensive and require supercomputer resources, see e.g. [11] and [24]. At the same time, many problems in the bioinformatics field may be solved by the resources that are widely available (on the contrast to supercomputers), namely, the desktop grid based systems [36]. The latter class of systems is mainly suitable for the specific class of problems which can be divided into many independent sub-problems (e.g. the so-called embarrassingly parallel applications), which is the case studied in the present paper.

Important ideas about causation and the study of the effect of the causes are given in fundamental work [14]. Accordingly, the causation studies developed in a few directions in accordance with mathematical structures used. One of these directions is based on graph-like structures for the causal relationships. In [28], such a structure is reconstructed as a tree (undirected acyclic graph) connecting the "visible variables" to the invisible (so-called hidden causes) using the conditional independence property.

In subsequent works, motivated by applications in social sciences and biology, the directed acyclic (DAG) [26] and directed cyclic graphs [33] were used as the models for causal relationships. Reconstruction of DAG by the so-called constraint-based method was suggested in the PC (Peter-Clark) algorithm proposed in [16] and further developed in [17] into the package `pcalg` for R language.

One of the key strengths of the PC algorithm implemented in the (recent versions of) `pcalg` package is related to the independence of the result on the order of the input, which is a very important feature for the causality studies [10]. A parallel version of the PC algorithm is presented in [20].

Finally, we mention the work [34] dedicated to comparative benchmarking of the causal relationship discovery algorithms which gives preference to the PC algorithm from `pcalg` package for medium and large size datasets.

2   Cloud Application for Causal Role Attribution

In this section we introduce and discuss the technical peculiarities of the approach used for the causal role attribution in the gene network expansion problem. The technical solution is novel, and is in fact a cloud-based web application built using the following key components (available as open source, or written earlier by the authors):

Shiny framework for building and running web applications having R language [32] interpreter server as a backend [9];

RBOINC R package as a software interface to the underlying distributed computing system [6];

BOINC framework for organizing and maintaining the desktop grid based distributed computing system [1,2];

VirtualBox virtualization software used to run the R tasks independently of the compute node operating system and in isolated environment [19];

pcalg R package for the PC algorithm for causal relationship discovery, used with parallel (multicore) computing option in fast (C++) implementation [17].

Hereafter we discuss a few details on implementation of this software combination.

2.1   Shiny Web Application

Here we describe the so-called *frontend* of the solution, which is a novel web application built using the Shiny framework. As the computational backend, the application uses the RBOINC package written and submitted earlier by the authors (see Sect. 2.2).

The Shiny framework based web application serves as the entrance point for the calculations of causal role attribution. Its aim is to simplify and automate the launch of calculations of the algorithm for detecting causal relationships between elements of the gene network. The application can be used for offering gene network causal studies as a cloud service over distributed computing backend. Selection of Shiny framework and R language is motivated by the fact that R ecosystem has a large number of packages in bioinformatics e.g. in Bioconductor repository [15].

At server side, the service uses gene expression level data for grapevine (the dataset is described in more details in Sect. 2.4). At user level, a batch of input files may be uploaded containing the so-called expansion lists in text format (the gene of interest and a list of related genes, see Sect. 2.4) to the server, and relative frequency threshold parameter (called *Frel* in the dataset, see [31] on the parameter interpretation) may be selected so as to find balance between the computation time and results accuracy. The filtered correlation matrix containing the genes of interest, as well as the corresponding R task, is then wrapped into a virtual machine and sent to BOINC-based desktop grid hosts using RBOINC package (see Sect. 2.2). The results may be obtained from the service using a text file of special format containing the batch descriptor. The results are downloaded in the form of a batch of files containing oriented graphs (in text format), each graph corresponding to one of the input files in the batch.

2.2   Distributed Computing Backend

The computations are based on BOINC-based desktop grid [2] and the (introduced earlier) RBOINC package that serves as the interface between R computations and BOINC infrastructure [7]. RBOINC software consists of 3 parts:

- `RBOINC.cl` R package that provides functions for interaction with BOINC server (necessary for batch processing of the workunits and results acquisition);
- *Server part* that is installed along with BOINC server and extends the application programming interfaces (API) of BOINC;
- Virtual machine (VM) build tools, i.e. collection of (shell) scripts that greatly simplify building the corresponding VM containing code and data sent to the BOINC hosts (computers used for distributed computing with BOINC).

`RBOINC.cl` is an R package that RBOINC users install into R environment in standard way. The package is capable of creating a batch of tasks directly from elements of R environment. Specifically, the user of the package indicates the function and the data to run the function over. The `RBOINC.cl` package then splits the data into several units and creates appropriate batch of workunits using BOINC API functions. Each such a workunit is in fact a virtual machine along with an appropriate R script and data to run the script over. As such, the computing environment is isolated from the host machine and may be effectively suspended and resumed during the busy times, without losing the intermediate results of computation.

A few other important functions of the package are:

- connection manipulation routines `create_connection` and `close_connection` interact with BOINC server using web API;
- workunit manipulation is performed by `create_jobs` and `cancel_jobs`;
- `update_jobs_status` is capable of results retrieval into R environment of the package user.

The server part is installed directly on the BOINC server and provides functions that are not implemented in BOINC or whose implementation in BOINC is too complicated to use. First of all, this is the creation of unique names, base templates for work units and results, as well as simple validator.

Finally, the VM build tools are a few scripts that greatly simplify creation of a VM image usable by RBOINC. In theory, they can be of help for any BOINC VirtualBox project. These scripts are compatible with Gentoo Linux (due to its small storage footprint and great flexibility) and allow one to automatically update most distribution packages to the latest version, install the necessary R packages and build a minimal[1] bootable disk for VM that can perform RBOINC jobs. To drastically reduce the size, root file system of the machine is compressed into a read-only squashfs image. Read/write access is achieved thought the use overlayfs with tmpfs (read-write file system in the random access memory). As part of RBOINC project, a preconfigured virtual machine is provided that contains the VM build scripts installed. Note that RBOINC is available at R-Forge [7].

2.3 Causal Attribution Algorithm

Several implementations of the famous PC algorithm were considered to be used as the functions to be sent using RBOINC.cl package to the hosts of the BOINC desktop grid. These implementations include the standard pcalg package [17] with various parameters and a parallel version from the ParallelPC package [21]. While authors of ParallelPC claim that the algorithm implements the stable version of the PC algorithm from pcalg package using parallel computing, an unwanted feature of the row order dependence was obtained in preliminary experiments with ParallelPC package, which is a significant disadvantage of the implementation (as opposed to the order-independent stable version in the pcalg package). As code study of the ParallelPC package version 1.2 (available at GitHub) has shown, the parameter solve.confl of the call of udag2pdagRelaxed function the original implementation of pc function in pcalg package version 2.7–8 (line 2241 of pcalg.R) was missing in the call of udag2pdagRelaxed in pc_parallel function of ParallelPC package (line 736 of ParallelPC.R). Modification of this line in accordance with original implementation in pcalg package makes the ParallelPC implementation order-independent.

A few preliminary runs were performed using the original R implementation of the algorithm in pcalg package, the (modified to be order-independent) version from ParallelPC as well as the pcalg version of pc function with skel.method="stable.fast" parameter. In all cases, the parameter solve.confl = TRUE was used so as to guarantee order independence on the input. The experiments were performed on a generic multicore desktop computer (as the typical desktop grid host). As a result, the following conclusions were drawn.

– Identical results were obtained using all three implementations.

---

[1] At least 330 MB, approximately 370 MB for this work due to additional packages installation. Required RAM dependents on task, minimum is 256 MB.

- The fastest implementation of the PC algorithm is the `pc` function from `pcalg` package with `skel.method="stable.fast"`, then goes the `ParallelPC` implementation, and finally, the original implementation from `pcalg` package with the parameter `skel.method = "stable"`.
- The fastest implementation, `pc` function with `skel.method="stable.fast"` parameter from `pcalg` package, allows to increase speedup by using multicore computation with a `numCores` parameter of the `pc` function.

The main reason for the high performance of the `stable.fast` implementation is probably the C++ language of implementation used behind the R wrapper of the function. Thus, a choice for validation experiment was made in favor of a standard package `pcalg` with `pc` function and `skel.method="stable.fast"` parameter together with `numCores=detectCores(logical = FALSE)-1` (with appropriate number of cores selected automatically).

2.4   Data Sources for Gene Network Expansion Problem: A Case Study

The presented approach can be applied for the transcriptomic datasets of various origin, and is not limited to the focus of the present paper, the *Vitis vinifera* dataset [25]. This, however, comes at a price of small modifications of the source code at the cloud service implementation step. In this section, we describe the dataset needed for validation case study which is of interest for this research.

In this case study, the dataset used at the server level of the cloud service uses the results of computation obtained by TN-Grid. The TN-Grid platform (https://gene.disi.unitn.it/test) is a joint project between the Institute of Materials for Electronics and Magnetism (IMEM) of the National Research Council of Italy (CNR), the Fondazione Edmund Mach (FEM), S. Michele all'Adige, Trento, Italy and the Department of Information Engineering and Computer Science (DISI) of the University of Trento, Italy. It uses the BOINC platform to provide advanced computational capabilities for local research activities. The project currently hosted on TN-Grid is called **gene@home** [4] and currently runs the OneGenE experiments [3], a graphical snapshot of the procedure is shown in Fig. 2 and how it is integrated into the scientific pipeline is illustrated on Fig. 3. In May 2023 the TN-Grid platform has $\approx$ 3500 registered users, $\approx$ 65000 registered hosts and a peak computational power of $\approx$ 33 TFLOPS.

The standard BOINC server architecture has been locally modified to suit the TN-Grid specific purposes, the overall picture of it and the changes are depicted in Fig. 1.

For this case study, the OneGenE experiment ran on a filtered and imputed version of the *Vitis vinifera* transcriptomic dataset [25]. The experiment computed the expansion, using the specific version of the PC algorithm known as PC-IM [5], of each of the 28013 genes it contains. The used PC-IM parameters were $tsize = 1000$, $iter = 2000$ and $\alpha = 0.05$. Any single (seed) gene expansion has been subdivided in 95 workunits (the computational unit sent to the volunteers) each one packed with 500 PC tiles. A cutoff of 2 has been applied

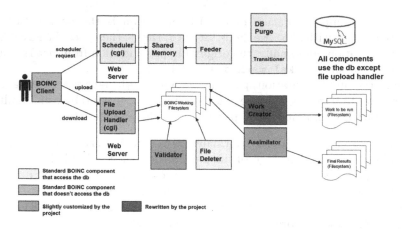

Fig. 1. BOINC on TN-Grid

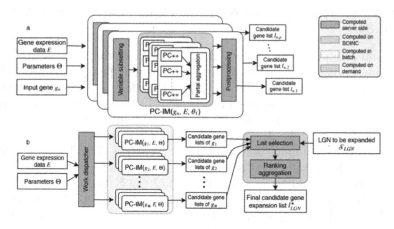

Fig. 2. The OneGenE experiment

to the output list, i.e. any couple of genes present less than two times has been discarded from the output. The initial post-processing phase combines together all the results of the 95 workunits discarding any couple which doesn't contain the seed gene thus obtaining an expansion list which is actually a undirected (the PC-IM algorithm doesn't keep track of the PC's separation set thus the direction of the interaction remains undefined) "star" graph centered on the seed gene, ordered by relative found frequency. These files with seed gene and expansion list were then used as the input files of the technical validation phase described in Sect. 3.

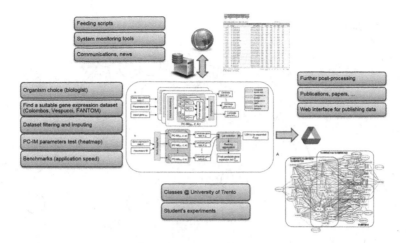

Fig. 3. The OneGenE scientific pipeline

3 Technical Validation Results

Hereafter we briefly report the preliminary results of validation of the approach to causal attribution described in Sect. 2. The grapevine dataset and its expansion lists described in Sect. 2.4 were used. For validation purpose, $N = 51$ expansion list L_i of the Ethylene Responsive Factor 45 (ERF45) gene was sent as a batch to the cloud service. ERF45 is a transcription factor, i.e. a gene which regulates other genes downstream in the cell signaling cascade. It is a marker of grapevine berry ripening onset, but its target genes have not been identified yet [22]. The computations for each such an expansion list were performed by the pc function from pcalg package with parameters maj.rule=TRUE, solve.confl=TRUE, u2pd ="relaxed" and skel.method="stable.fast". To enable multicore parallelism at the host level, in the function pc the parameter numCores = detectCores(logical = FALSE)-1 was used as well. The results were obtained from the cloud service and summarized as follows: N directed acyclic graphs (DAG) $G_i = \langle V_i, E_i \rangle$, $i = 1, \ldots, N$, were received as the results of computation. These graphs were then merged into a single graph $G = \langle V, E \rangle$ in the following steps.

1. All the edge data from the sets $E_i, i = 1, \ldots, N$, were aggregated by summing up the number of times each edge, say, $e \in E = \{(a, b), a \leq b \in V := \cup_{i=1}^{N} V_i\}$ was having (say) *right*, $r(e)$, *left*, $l(e)$, or *no*, $u(e)$ direction (the latter option in fact means that the algorithm wasn't able to find evidence on selecting a specific direction, which can be interpreted as selecting both directions with equal probability).
2. For each edge e, two attributes were added. The *weight*, $w(e)$ was defined as the number of times the edge (of any direction) was found in the results, divided by the number of times the pair of vertices of the edge was found in the input files,

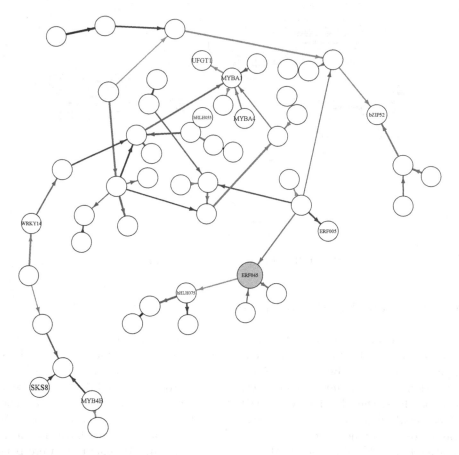

Fig. 4. Validation results of the experiment with causal attribution of the graph composed after processing of ERF45 gene expansion lists. The gene ERF45 is highlighted with gray color. The edge thickness is obtained in (1) and intensity of the gray color is obtained as in (2), whereas the direction of the edge is selected as the majority from the possible directions obtained in the experiments.

$$w(e) = \frac{r(e) + l(e) + u(e)}{\sum_{i=1}^{N} 1_{e \in L_i}} \leq 1, \quad e \in E. \tag{1}$$

The *color intensity* $c(e)$ was then obtained as the number of times the specific orientation was obtained, compared to the overall number of times the edge was present in the output,

$$c(e) = \frac{\max(r(e), l(e))}{r(e) + l(e) + u(e)} \leq 1, \quad e \in E. \tag{2}$$

3. Finally, the edges were filtered so as to have $w(e) \geq 0.5$ and $c(e) \geq 0.5$, and the largest connected component of the graph $G = \langle V, E \rangle$ was obtained.

The resulting graph G was depicted on the Fig. 4 with `igraph` library using the weight as $w(e)$ in terms of edge thickness, color as $c(e)$ in terms of gray color intensity, and direction selected according to $\max(r(e), l(e))$ (in obvious notation). The nodes are named using gene symbol (if available). The specific codes were omitted in attempt to make the graph more readable. The so-called Davidson–Harel algorithm for locating the nodes was used, by the parameter layout_with_dh of the igraph function. Such an approach allows to build and depict the graph as the results of the batch computation in an automatic way.

This analysis is quite valuable as it produces a prediction of causal relationships between gene pairs that can be tested experimentally. For ERF45, we identified two target genes, one of these bHLH075, has been already characterized for being an important regulator of ripening [12], and two genes potentially regulating ERF45, which are two transcription factors themselves and are interesting candidates for further studies. Nonetheless, there are still some critical issues that we need to further investigate to refine this prediction, but they concern the post-processing part. In particular, we need to better understand the factors affecting the direction, as it is not always conserved, to improve the final estimation.

4   Conclusion

Regulatory networks applying causal inference to transcriptomic data are still quite a rare resource for plant biologists, so the gene networks elaborated by this approach on a genome-wide scale represent a powerful tool that can support and boost molecular plant science. Parallel and distributed computing using BOINC platform proved to be a successful method to speed up research and manage big project in a reasonable time. Large scale experiments are planned to be performed.

We note that the software solution is currently under development, so there is no public repository available yet. However, it is planned to avail the solution to the community, and before that the setup can be repeated by the researchers familiar with the appropriate technologies independently using the description presented in this paper. As a technical perspective, we can also mention that the setup presented in this paper is not limited to the PC algorithm, and so one of the future directions may be in comparison of various causal attribution algorithms (see e.g. [34]) on the basis of the dataset used.

Acknowledgements. The research described in this publication was made possible in part by R&D Support Program for undergraduate and graduate students and post-doctoral researchers of PetrSU, funded by the Government of the Republic of Karelia.

References

1. Anderson, D.: BOINC: a platform for volunteer computing. J. Grid Comput. **18**, 99–122 (2020)
2. Anderson, D.P.: BOINC: a system for public-resource computing and storage. In: Proceedings of the 5th IEEE/ACM International Workshop on Grid Computing, GRID 2004, Washington, DC, USA, pp. 4–10. IEEE Computer Society (2004). https://doi.org/10.1109/GRID.2004.14

3. Asnicar, F., Masera, L., Pistore, D., Valentini, S., Cavecchia, V., Blanzieri, E.: OneGenE: regulatory gene network expansion via distributed volunteer computing on BOINC. In: 2019 27th Euromicro International Conference on Parallel, Distributed and Network-Based Processing (PDP), Pavia, Italy, pp. 315–322. IEEE (2019). https://ieeexplore.ieee.org/document/8671629/

4. Asnicar, F., et al.: TN-Grid and gene@home project: volunteer computing for bioinformatics. In: Proceedings of the Second International Conference BOINC-based High Performance Computing: Fundamental Research and Development (BOINC:FAST 2015), vol. 1502, pp. 1–15. CEUR-WS (2015). https://ceur-ws.org/Vol-1502/paper1.pdf

5. Asnicar, F., et al.: TN-grid and gene@home project: volunteer computing for bioinformatics. In: Proceedings of the Second International Conference BOINC-based High Performance Computing: Fundamental Research and Development (BOINC:FAST 2015). No. 1502 in CEUR Workshop Proceedings, Aachen (2015). https://ceur-ws.org/Vol-1502

6. Astafiev, S.N., Rumyantsev, A.S.: Distributed computing of R applications using RBOINC package with applications to parallel discrete event simulation. In: Vishnevskiy, V.M., Samouylov, K.E., Kozyrev, D.V. (eds.) Distributed Computer and Communication Networks. CCIS, vol. 1552, pp. 396–407. Springer, Cham (2022). https://doi.org/10.1007/978-3-030-97110-6_31. iSSN 1865–0937

7. Astafiev, S., Rumyantsev, A.: R-Forge: RBOINC (2022). https://r-forge.r-project.org/projects/rboinc/

8. Blanzieri, E., et al.: A computing system for discovering causal relationships among human genes to improve drug repositioning. IEEE Trans. Emerg. Top. Comput. **9**(4), 1667–1682 (2021). https://ieeexplore.ieee.org/document/9224179/

9. Chang, W., et al.: shiny: Web Application Framework for R (2023). https://shiny.rstudio.com/. r package version 1.7.4.9002

10. Colombo, D., Maathuis, M.H.: Order-independent constraint-based causal structure learning. J. Mach. Learn. Res. **15**, 3741–3782 (2014)

11. Dumancas, G.G.: Applications of supercomputers in sequence analysis and genome annotation. In: Advances in Systems Analysis, Software Engineering, and High Performance Computing, pp. 149–175. IGI Global (2015). https://doi.org/10.4018/978-1-4666-7461-5.ch006

12. Fasoli, M., et al.: Timing and order of the molecular events marking the onset of berry ripening in grapevine. Plant Physiol. **178**(3), 1187–1206 (2018). https://academic.oup.com/plphys/article/178/3/1187-1206/6116656

13. Hernan, M.A., Robins, J.M.: Causal Inference: What If. Chapman & Hall/CRC, Boca Raton (2020). https://www.hsph.harvard.edu/miguel-hernan/causal-inference-book/

14. Holland, P.W.: Statistics and causal inference. J. Am. Stat. Assoc. **81**(396), 945–960 (1986). https://www.tandfonline.com/doi/abs/10.1080/01621459.1986.10478354

15. Huber, W., et al.: Orchestrating high-throughput genomic analysis with Bioconductor. Nat. Methods **12**(2), 115–121 (2015). https://www.nature.com/nmeth/journal/v12/n2/full/nmeth.3252.html

16. Kalisch, M., Bühlmann, P.: Estimating high-dimensional directed acyclic graphs with the PC-algorithm. J. Mach. Learn. Res. **8**, 613–636 (2007)

17. Kalisch, M., Mächler, M., Colombo, D., Maathuis, M.H., Bühlmann, P.: Causal inference using graphical models with the R Package pcalg. J. Stat. Softw. **47**(11) (2012). https://www.jstatsoft.org/v47/i11/

18. Kenny, D.A.: Correlation and Causality. Wiley, New York (1979)
19. Khan, R., AlHarbi, N., AlGhamdi, G., Berriche, L.: Virtualization software security: oracle VM VirtualBox. In: 2022 Fifth International Conference of Women in Data Science at Prince Sultan University (WiDS PSU), Riyadh, Saudi Arabia, pp. 58–60. IEEE (2022). https://ieeexplore.ieee.org/document/9764794/
20. Le, T.D., Hoang, T., Li, J., Liu, L., Liu, H., Hu, S.: A fast PC algorithm for high dimensional causal discovery with multi-core PCs. IEEE/ACM Trans. Comput. Biol. Bioinform. **16**(5), 1483–1495 (2019). https://ieeexplore.ieee.org/document/7513439/
21. Le, T.D., Xu, T., Liu, L., Shu, H., Hoang, T., Li, J.: ParallelPC: an R package for efficient causal exploration in genomic data. In: Ganji, M., Rashidi, L., Fung, B.C.M., Wang, C. (eds.) PAKDD 2018. LNCS (LNAI), vol. 11154, pp. 207–218. Springer, Cham (2018). https://doi.org/10.1007/978-3-030-04503-6_22
22. Leida, C., et al.: Insights into the role of the berry-specific ethylene responsive factor VviERF045. Front. Plant Sci. **7** (2016). https://journal.frontiersin.org/article/10.3389/fpls.2016.01793/full
23. Li, J., Liu, L., Le, T.D.: Practical Approaches to Causal Relationship Exploration. SECE. Springer, Cham (2015). https://doi.org/10.1007/978-3-319-14433-7
24. Maples, R., Ramasahayam, S., Dumancas, G.G.: Supercomputers in modeling of biological systems. In: Advances in Systems Analysis, Software Engineering, and High Performance Computing, pp. 201–222. IGI Global (2015). https://doi.org/10.4018/978-1-4666-7461-5.ch008
25. Moretto, M., et al.: VESPUCCI: exploring patterns of gene expression in grapevine. Front. Plant Sci. **7** (2016). https://journal.frontiersin.org/Article/10.3389/fpls.2016.00633/abstract
26. Pearl, J.: Probabilistic Reasoning in Intelligent Systems: Networks of Plausible Inference. Morgan Kaufmann Series in Representation and Reasoning, Elsevier Science (1988)
27. Pearl, J.: Causality: Models, Reasoning, and Inference, 2nd edn. Cambridge University Press (2009). https://www.cambridge.org/core/product/identifier/9780511803161/type/book
28. Pearl, J., Tarsi, M.: Structuring causal trees. J. Complex. **2**(1), 60–77 (1986). https://linkinghub.elsevier.com/retrieve/pii/0885064X86900233
29. Peters, J., Janzing, D., Schölkopf, B.: Elements of Causal Inference: Foundations and Learning Algorithms. Adaptive Computation and Machine Learning Series. The MIT Press, Cambridge (2017)
30. Pilati, S., et al.: Vitis OneGenE: a causality-based approach to generate gene networks in vitis vinifera sheds light on the laccase and dirigent gene families. Biomolecules **11**(12), 1744 (2021). https://doi.org/10.3390/biom11121744
31. Pilati, S., et al.: Vitis OneGenE: 1 causality-based approach to generate gene networks in vitis vinifera sheds light on the laccase and dirigent gene families. Biomolecules **11**(12) (2021). https://www.mdpi.com/2218-273X/11/12/1744
32. R Core Team: R: A Language and Environment for Statistical Computing. R Foundation for Statistical Computing, Vienna, Austria (2021). https://www.R-project.org
33. Richardson, T.: A discovery algorithm for directed cyclic graphs. In: Proceedings of the Twelfth International Conference on Uncertainty in Artificial Intelligence, UAI 1996, pp. 454–461. Morgan Kaufmann Publishers Inc., San Francisco (1996). Event-place: Portland, OR

34. Singh, K., Gupta, G., Tewari, V., Shroff, G.: Comparative benchmarking of causal discovery algorithms. In: Proceedings of the ACM India Joint International Conference on Data Science and Management of Data, Goa India, pp. 46–56. ACM (2018). https://dl.acm.org/doi/10.1145/3152494.3152499
35. Spirtes, P., Glymour, C.N., Scheines, R.: Causation, Prediction, and Search. Adaptive Computation and Machine Learning, 2nd edn. MIT Press, Cambridge (2000)
36. Talbi, E.G., Zomaya, A.Y. (eds.): Grid Computing for Bioinformatics and Computational Biology. Wiley, Hoboken (2007). https://doi.org/10.1002/9780470191637

HiTViSc: High-Throughput Virtual Screening as a Service

Natalia Nikitina[1]([✉])[iD] and Evgeny Ivashko[1,2][iD]

[1] Institute of Applied Mathematical Research, Karelian Research Center of the Russian Academy of Sciences, 185910 Petrozavodsk, Russia
nikitina@krc.karelia.ru
[2] Petrozavodsk State University, 185035 Petrozavodsk, Russia

Abstract. High-performance and high-throughput computing play an important role in drug development and, in particular, in solving the computationally intensive problem of virtual screening. The variety and complexity of tools require technical knowledge for selection, setup and usage of the computational platform. There is need for ready solutions and services to simplify the process. With low cost and high scalability, Desktop Grid systems can significantly expand the computational capacity available for a virtual screening. This paper describes High-Throughput Virtual Screening as a Service (HiTViSc): we present three logical levels of operation (computational, virtual screening and user level), the user workflows related to virtual screening, resource administration and visualization and analysis of results, and the multi-user access. The novelty of the work is related to implementation of the Desktop Grid as a Service concept. In particular, comparing to other cloud-based virtual screening services, we use Desktop Grid resources to implement computationally intensive work. Comparing to umbrella Desktop Grid projects, the users of HiTViSc can be both consumers and providers of computing resources at the same time, and employ additional steps of virtual screening based on supportive utilities provided by HiTViSc.

Keywords: High-throughput computing · Desktop Grid · Volunteer computing · Cloud computing · Virtual screening · BOINC · HiTViSc

1 Introduction

Drug development is a complex, multi-stage, resource-consuming process and is assisted by *in silico* methods also known as CADD (computer-aided drug design) [21, 23]. Recent literature emphasizes the following main directions of contemporary development of CADD technologies:

- Enhancement of the iterative exchange with a laboratory, experimental evaluation and validation of *in silico* results;
- Application of new computational technologies and expansion of their scope to reduce time and money needed for drug development;

V. Malyshkin (Ed.): PaCT 2023, LNCS 14098, pp. 83–92, 2023.
https://doi.org/10.1007/978-3-031-41673-6_7

– Enhancement of mathematical and computer models of biological processes.

High-performance computing (HPC) technologies play an important role in solving these challenges, as they allow researchers to process large amounts of data and implement complex simulations [14]. As an alternative, one may use high-throughput computing (HTC) to solve large numbers of loosely-coupled computational tasks, which arise in multiple stages of CADD such as virtual screening, molecular dynamics simulations with different parameters, modeling protein folding, etc. Conventional HTC systems are clusters, commercial clouds and Grids. In our work, we concentrate on another type of HTC systems: Desktop Grids. Desktop Grid is a system using idle time of non-dedicated geographically distributed computing nodes (typically, desktop computers) connected to the central server by the Internet or a local access network. Such systems were proposed in 1980s s and have been used by many of the world's leading research institutions for large-scale computational projects (e.g., Washington University: Rosetta@home [22], Folding@home [26]; CERN: LHC@home [8]; University of Oxford: Climateprediction.net [5]).

An example of a computationally intensive problem well fit to be solved on Desktop Grids is the virtual screening, an *in silico* alternative to high-throughput screening. It is a computational technique allowing to process large libraries of small chemical compounds and obtain the molecules with high predicted binding activity against a specified therapeutic target. The selection is made basing on the results of molecular docking of each small compound to the target.

This paper is organized as follows. In Sect. 2, we review related work on the usage of HPC/HTC for virtual screening. In Sect. 3, we present the High-Throughput Virtual Screening as a Service. In Sect. 4, we describe the physical setup of the system. Finally, Sect. 5 concludes the paper.

2 Related Work

Virtual screening is one of the most computationally intensive stages of drug development. Despite the fact that molecular docking of a ligand to a target is fast, the volumes of ligand libraries lead to the need for HPC/HTC tools for virtual screening. In [12], the authors review the variety of HPC options when performing virtual screening with AutoDock Vina software and provide important observations on computational efficiency and reproducibility.

In a recent review [16], the authors consider implementations of parallelization algorithms for virtual screening on HPC systems. In particular, they emphasize the computational complexity and discuss parallelization strategies at all steps of virtual screening. Another recent review [25] focuses on molecular docking on supercomputers and illustrates the demand for HPC systems for virtual screening. The methodology of GPU acceleration with appropriate numerical optimization for virtual screening has been described in [6] on the example of the Summit supercomputer.

This illustrates that HPC technologies provide biochemist scientists with computational systems of various scales, efficiencies and costs. But on the other hand, the variety and complexity of HPC/HTC tools require technical knowledge for selection, setup and usage of a platform. When the platform has been selected and setup, one needs to orchestrate many software programs and data processing algorithms in the process of drug development. This leads to the need for ready solutions and services to simplify the process. Recent overviews of such services based on cloud computing are given, for example, in [20,24]. A service based on the Chinese National Grid CNGrid is described in [28]; based on the supercomputer Tianhe-2 – in [15].

There are a number of commercial solutions that provide a full cycle of virtual screening based on cloud computational resources – Virtual Screening as a Service [1,7,13]. Such software solutions are proprietary, and within their framework it is difficult (or impossible) to implement own algorithms for library preparation and the whole computational experiment, and the amount of available computational resources is limited.

At the same time, Desktop Grid computing gives a potentially large computational capacity at a low cost. This is important because research groups (especially of a small/medium size) typically do not have immediate, on-demand access to supercomputing resources. Desktop Grids complement HPC tools when performing virtual screening (see, for instance, [27]). A range of research works investigate various combinations of Desktop Grids and cloud computing, aiming to gain advantages of both concepts (see [10] for an overview).

3 High-Throughput Virtual Screening as a Service

The aim of this work is to develop a computing service that, on the one hand, offers users a convenient web interface for virtual screening, visualization and a primary analysis of results, and on the other hand, provides an HTC tool for virtual screening using distributed computing based on Desktop Grids, including the feature of administration of available computational resources.

3.1 Logical View

High-Throughput Virtual Screening as a Service system implements the concept of Desktop Grid as a Service [10], which provides the user with scalable Desktop Grid resources in the form of a specialized cloud service.

From the logical viewpoint, High-Throughput Virtual Screening as a Service has three implementation levels (see Fig. 1):

1. The first level - computational - provides high-performance computations basing on resources of a Desktop Grid. At this level, a Desktop Grid server implements the functions of task generation, communication with computing nodes (task assignment, transfer of input data and computational applications) and results accounting. This level can be implemented on the basis of an existing Desktop Grid software platform, for example, BOINC.

2. The second level - virtual screening - implements special virtual screening functions, including uploading a target file, a ligand database and a computational application, parameter settings, a protocol for molecular docking and results processing, and the selection of external applications for analysis and visualization of results.
3. The third level - user level - provides the graphical user interface of the cloud application, including interfaces and settings for management of the virtual screening, progress visualization, results visualization and analysis, as well as administration of computational resources.

The computing nodes are Desktop Grid clients that provide distributed computational resources.

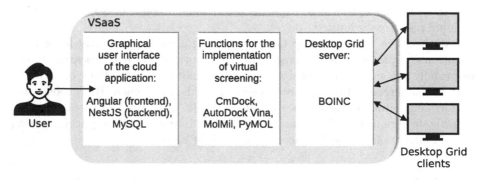

Fig. 1. High-Throughput Virtual Screening as a Service: logical levels and the main software tools used in the implementation of each level.

3.2   Workflows

The primary user of the system is a scientist (alone or in a group) conducting virtual screening within three workflows: 1) execution of a computational experiment, 2) administration of computational resources, 3) expert analysis of the results.

1. The computational experiment is performed as follows:
   (a) *Load the target.* PDB format is widely used to describe the 3D structure of the target for molecular docking. This format is used in the leading open database of structural protein models RCSB Protein Data Bank (RCSB PDB) [3]. In the current practice of virtual screening, the user uploads a PDB file or specifies a PDB identifier in the RCSB PDB database. Then the file is prepared for molecular docking by software utilities and/or manually.
   (b) *Select the ligand library.* Preparation of the ligand library is an expert task, the result of which affects the quality of virtual screening. File format depends on the selected molecular docking program. For example, SDF format is used for CmDock, PDBQT format is used for AutoDock Vina,

etc. The initial selection is made among the libraries available on the system server, accessible via online interfaces or uploaded by the user. The library is then filtered according to the user's requirements. The prepared ligand library is stored on the server and is an important part of a specific computational experiment of virtual screening. It is also possible to specify a set of reference ligands that the user will need to prepare the target and/or analyze the results of virtual screening.

(c) *Select the molecular docking program.* There are many software (such as the already mentioned CmDock, AutoDock Vina) that differ both conceptually and technically (settings format, file formats, etc.). In current practice, the entire virtual screening process is designed for a software available on the server or locally.

(d) *Configure the docking protocol.* The type and format of the parameters depend on the specific molecular docking program. For example, for CmDock, these are 3D coordinates of the center of the binding site and the size of the docking area, the number of repeats of molecular docking, various restrictions and filters[1]. The system interface provides a default protocol with the ability to edit settings.

(e) *Set up the protocol of the computational experiment.* The protocol determines the criteria for hits selection and for stopping the computational experiment.

 – The criterion for hits selection is the binding energy value to consider the ligands as hits. It may be a threshold value, ligand efficiency (binding energy normalized by the number of atoms) or a more complex selection criterion.
 – A criterion for stopping a computational experiment may be, for example, a given number of tested molecules, a given number of hits, or a given proportion of hits in a fixed number of tested molecules.

(f) *Select computational resources.* To perform a virtual screening, the user has options for selecting computational resources. At the same time, in accordance with the concept of Desktop Grid, the basis is the resources of non-dedicated computers, including those invited to the system by the user. Computations within the framework of the experiment can be of one of three types:

 i. testing: to verify the correctness of the experiment settings,
 ii. public: the results of the experiment, including the target, settings and results of virtual screening are available to a wide audience,
 iii. private: the results of the experiment are available only to the user and his partners.

(g) *Start or stop a computational experiment.*

(h) *Display the progress of the computational experiment.* From the point of view of the experiment dynamics, the following parameters are central:

 i. the proportion of processed molecules from a given library,

---

[1] https://gitlab.com/Jukic/cmdock/-/blob/master/docs/reference-guide/docking-protocol.rst.

ii. the number of hits found,

iii. library coverage (for example, the number of clusters),

iv. a forecast of the completion time of the computational experiment.

2. Administration of computational resources. The service concept assumes three types of computational resources that can be used in virtual screening as Desktop Grid computing nodes:

   (a) Test environment. A limited amount of highly available and reliable computational resources used to test parameters on a small amount of computations.

   (b) Computational resources of the system provided for general use by volunteers and other users of High-Throughput Virtual Screening as a Service and used to perform experiments.

   (c) Own and partners' computational resources connected to the system, which can be used for own experiments, partner projects or provided for general use.

   It is important to take into account the computational resources actually used. One of the possible policies of the service may be that the priority of a particular experiment decreases or increases depending on how many resources does the user provide to other projects. At the same time, the connection of "own" resources is carried out using a special connection link.

3. Analysis of the results of virtual screening. The results of the virtual screening are a list of hits ordered by estimates of their binding energy to the target. For each hit, an estimate of the binding energy and a specific conformation corresponding to this estimate and representing a three-dimensional spatial model of the molecule are stored. In practice, the most often used methods for the analysis and visualization of the results are the principal component analysis and clustering. They are performed by third-party programs (such as R, Clustvis, etc.) that implement suitable algorithms.

   Another basic method for analyzing the results of virtual screening is visualization. One uses 2D visualization in the form of points projected onto a plane (for visual evaluation of clustering and chemical diversity) and 3D visualization in the form of molecular structures docked to the target (for visual evaluation of the results of molecular docking and comparison with reference ligands, if any).

   Finally, the function of uploading data filtered by any criteria is a universal opportunity to use third-party independent programs to analyze the results.

3.3 Multi-user Access

High-Throughput Virtual Screening as a Service is a scalable multi-user system. This means that within the framework of the service, many users can independently conduct their computational experiments. At the same time, in general, each user has access to the entire pool of computational resources which is shared among users by the task scheduler.

A multi-user approach is also used within the computational project: the owner (founder) of the project can provide other participants with more or less limited access to the project settings, computational resources and results. Such a division of user roles corresponds to the practices adopted in scientific research. Each user is provided with the following features:

1. Access to an account for performing computational experiments, storing and analyzing the results.
2. Use of computational resources.
3. Sharing with the other users the results of computational experiments according to one of three access options: the results are available to all users of the system, to a certain group of users, or to the author only.

The main considered operational mode implies that all generated data (prepared targets and ligand libraries, found hits, binding energy estimates, etc.) can be used by a wide range of users for further expert analysis with the preservation of the authorship.

At the same time, an open library of results is, on the one hand, the basis for attracting volunteers to the project to provide computational resources (as this is a common motivation in volunteer computing), and on the other hand, a contribution to the overall scientific progress in the field of bioinformatics.

The described multi-user environment is different from standard cloud services and the BOINC system in that users of HiTViSc system are both consumers and providers of computational resources. But unlike BOINC umbrella projects where users can run their own subprojects and volunteers compute for them, HiTViSc system offers a virtually unlimited flexibility of the experiment setup. Also, there is no need for a public presentation of the project, with an option of running a private instance of the whole system which is useful at the stage of preliminary experiments that might take months in a real setting.

4 System Setup

As described above, the proposed system operates on three logical levels. On the first level, we employ BOINC software: an open source platform for Desktop Grid computing with a client-server architecture. BOINC server consists of a number of parallel services sharing a database (see [2] for detailed architecture). BOINC client connects to the server to request tasks, performs computations and sends the results back to the server. The server validates and assimilates the results, aggregating them to obtain hits of virtual screening.

Here, the clients are desktops and other types of computers that can be provided by the user, volunteers and other individuals. The total computational performance depends on the project needs. Practice shows that it is possible to gather several tens of computers for a short-term, local experiment [11] or several thousand computers for a long-term public project on virtual screening [19].

On the second level, the management of virtual screening is implemented at a dedicated server, which can be shared with BOINC, by virtual screening-specific applications like molecular docking software (CmDock, AutoDock Vina), third-party molecular visualization systems (Molmil, PyMOL), R scripts for results analysis and visualization, etc. It holds the functionality of the cloud application.

On the third level, the graphical user interface is implemented as a web portal built on Angular, NodeJS and MySQL. The main software tools used for implementation of each level are shown in Fig. 1.

At the moment, the implementation of HiTViSc system is in progress. In the design and development of the system, we base on a large series of computational experiments previously conducted locally [11] in a BOINC-based Desktop Grid using the molecular docking application AutoDock Vina and in a public BOINC project Sidock@home [19] using CmDock. The experiments embraced the whole cycle of virtual screening in a Desktop Grid environment. The majority of BOINC clients have been desktop computers, but overall, the hardware varied from dedicated servers to routing devices. The experiments in SiDock@home involved up to 10,000 heterogeneous computers simultaneously. It took on average 1.5 months for virtual screening of 1-billion ligands library (an analogue is the whole ZINC dataset [9]) against one therapeutical target. With further optimization of the workflow [17], expected processing time of one target is going to be about 3 weeks. The statistics gathered during the experiments allowed to design the database and system environment of HiTViSc system.

5   Conclusion

High-performance computing plays a significant role in conducting modern fundamental and applied research and development. To date, many problems cannot be solved without the use of HPC/HTC systems. At the same time, to solve a specific computationally complex problem, an appropriate tool should be selected that allows one to find a solution effectively.

Desktop Grid systems have proven themselves well in solving the problem of virtual screening, providing high performance and scalability. However, the process of deploying Desktop Grid and maintaining the computing process is usually technically difficult for scientists (this reason, in particular, affects the number of existing relevant volunteer computing projects). In this work, we suggest a solution to this problem: High-Throughput Virtual Screening as a Service, a cloud-based virtual screening service based on Desktop Grid, built on the principles of Desktop Grid as a Service.

The presented paper describes High-Throughput Virtual Screening as a Service (HiTViSc): we present three logical levels of operation (computational, virtual screening and user level), the user workflows related to virtual screening, resource administration and visualization and analysis of results, and the principle of multi-user access. Further work is related to the development of architecture, design and implementation of the High-Throughput Virtual Screening as a Service. In particular, an important direction of work is development of new efficient problem-specific task scheduling algorithms for the exploration of ligand library, analysis of hits and iterative process of virtual screening.

It is important to note that the proposed architecture is not restricted to virtual screening only. The developed approach is applicable to solving any computationally-intensive problems that may be implemented on the basis of

HTC and, in particular, Desktop Grids. Specifically, these are the problems belonging to the class of Bag-of-tasks (BoT) which are present in many applications of science and technology. A modular, flexible architecture of HiTViSc allows to adapt it for different computational problems.

Moreover, problem-specific mathematical models and algorithms that will be implemented on the second logical level of the system (as presented in Subsect. 3.1) are not restricted to for virtual screening either, and can be expanded to a broad range of problems. For example, the task scheduling algorithm for fast discovery of diverse results, earlier proposed by the authors [18], can be efficient for exploring parameter spaces, and has been applied in solving a problem of parameter identification of a hydride decomposition model [4].

References

1. Blaze Cloud from Cresset. https://www.cresset-group.com/products/blaze/#blaze-cloud. Accessed 31 Jan 2023
2. Anderson, D.P.: BOINC: a platform for volunteer computing. J. Grid Comput. **18**(1), 99–122 (2020)
3. Berman, H.M., et al.: The protein data bank. Nucl. Acids Res. **28**(1), 235–242 (2000). https://doi.org/10.1093/nar/28.1.235
4. Chernov, I.: Effective scanning of parameter space in a desktop grid for identification of a hydride decomposition model. Program Syst. Theory Appl. **9**(4(39)), 53–68 (2018). https://doi.org/10.25209/2079-3316-2018-9-4-53-68
5. Climateprediction.net | the world's largest climate modelling experiment for the 21st century. https://www.climateprediction.net. Accessed 31 Mar 2023
6. Glaser, J., et al.: High-throughput virtual laboratory for drug discovery using massive datasets. Int. J. High Perform. Comput. Appl. **35**(5), 452–468 (2021)
7. Hawkins, P.: Virtual Screening At Ultra-Large Scale: 1.5 Billion And Counting - Webinars. https://www.healthtech.com/openeye-scientific-virtual-screening-at-ultra-large-scale/. Accessed 31 Jan 2023
8. Home | LHC@home. https://lhcathome.web.cern.ch. Accessed 31 Mar 2023
9. Irwin, J.J., Sterling, T., Mysinger, M.M., Bolstad, E.S., Coleman, R.G.: Zinc: a free tool to discover chemistry for biology. J. Chem. Inf. Model. **52**(7), 1757–1768 (2012). https://doi.org/10.1021/ci3001277
10. Ivashko, E.: Desktop Grid as a service concept. In: Voevodin, V., Sobolev, S., Yakobovskiy, M., Shagaliev, R. (eds.) Supercomputing: 8th Russian Supercomputing Days, RuSCDays 2022. LNCS, vol. 13708, pp. 632–643. Springer, Cham (2022). https://doi.org/10.1007/978-3-031-22941-1_46
11. Ivashko, E.E., Nikitina, N.N., Möller, S.: High-performance virtual screening in a BOINC-based Enterprise Desktop Grid. Vestnik Yuzhno-Ural'skogo Gosudarstvennogo Universiteta. Seriya Vychislitelnaya Matematika i Informatika **4**(1), 57–63 (2015)
12. Jaghoori, M.M., Bleijlevens, B., Olabarriaga, S.D.: 1001 ways to run AutoDock Vina for virtual screening. J. Comput. Aided Mol. Des. **30**, 237–249 (2016)
13. Krasoulis, A., Antonopoulos, N., Pitsikalis, V., Theodorakis, S.: DENVIS: scalable and high-throughput virtual screening using graph neural networks with atomic and surface protein pocket features. J. Chem. Inf. Model. **62**(19), 4642–4659 (2022)
14. Liu, T., et al.: Applying high-performance computing in drug discovery and molecular simulation. Natl. Sci. Rev. **3**(1), 49–63 (2016)

15. Mo, Q., Xu, Z., Yan, H., Chen, P., Lu, Y.: VSTH: a user-friendly web server for structure-based virtual screening on Tianhe-2. Bioinformatics **39**(1), btac740 (2023)
16. Murugan, N.A., Podobas, A., Gadioli, D., Vitali, E., Palermo, G., Markidis, S.: A review on parallel virtual screening softwares for high-performance computers. Pharmaceuticals **15**(1), 63 (2022)
17. Nikitina, N., Ivashko, E.: Optimization of the workflow in a BOINC-based Desktop Grid for virtual drug screening. In: Voevodin, V., Sobolev, S., Yakobovskiy, M., Shagaliev, R. (eds.) Supercomputing, RuSCDays 2022. LNCS, vol. 13708, pp. 686–698. Springer, Cham (2022). https://doi.org/10.1007/978-3-031-22941-1_50
18. Nikitina, N., Ivashko, E., Tchernykh, A.: Congestion game scheduling implementation for high-throughput virtual drug screening using BOINC-based Desktop Grid. In: Malyshkin, V. (ed.) PaCT 2017. LNCS, vol. 10421, pp. 480–491. Springer, Cham (2017). https://doi.org/10.1007/978-3-319-62932-2_46
19. Nikitina, N., Manzyuk, M., Podlipnik, Č, Jukić, M.: Volunteer computing project SiDock@home for virtual drug screening against SARS-CoV-2. In: Byrski, A., Czachórski, T., Gelenbe, E., Grochla, K., Murayama, Y. (eds.) ANTICOVID 2021. IAICT, vol. 616, pp. 23–34. Springer, Cham (2021). https://doi.org/10.1007/978-3-030-86582-5_3
20. Olğaç, A., Türe, A., Olğaç, S., Möller, S.: Cloud-based high throughput virtual screening in novel drug discovery. In: Kołodziej, J., González-Vélez, H. (eds.) High-Performance Modelling and Simulation for Big Data Applications. LNCS, vol. 11400, pp. 250–278. Springer, Cham (2019). https://doi.org/10.1007/978-3-030-16272-6_9
21. Prieto-Martínez, F.D., López-López, E., Juárez-Mercado, K.E., Medina-Franco, J.L.: Computational drug design methods-current and future perspectives. In: Silico Drug Design, pp. 19–44 (2019)
22. Rosetta@home. https://boinc.bakerlab.org. Accessed 31 Mar 2023
23. Sabe, V.T., et al.: Current trends in computer aided drug design and a highlight of drugs discovered via computational techniques: a review. Eur. J. Med. Chem. **224**, 113705 (2021)
24. Singh, N., Chaput, L., Villoutreix, B.O.: Virtual screening web servers: designing chemical probes and drug candidates in the cyberspace. Brief. Bioinform. **22**(2), 1790–1818 (2021)
25. Sulimov, A.V., Kutov, D.C., Sulimov, V.B.: Supercomputer docking. Supercomput. Front. Innov. **6**(3), 26–50 (2019)
26. Together We Are Powerful - Folding@home. https://foldingathome.org. Accessed 31 Mar 2023
27. Zhang, B., D'Erasmo, M.P., Murelli, R.P., Gallicchio, E.: Free energy-based virtual screening and optimization of RNase H inhibitors of HIV-1 reverse transcriptase. ACS Omega **1**(3), 435–447 (2016)
28. Zhang, B., Li, H., Yu, K., Jin, Z.: Molecular docking-based computational platform for high-throughput virtual screening. CCF Trans. High Perform. Comput. 1–12 (2022)

Expanding the Cellular Automata Topologies Library for Parallel Implementation of Synchronous Cellular Automata

Yuri Medvedev[1,2(✉)] ⓘ, Sergey Kireev[1,2] ⓘ, and Yulia Trubitsyna[2]

[1] Institute of Computational Mathematics and Mathematical Geophysics SB RAS, Novosibirsk, Russia
[2] Novosibirsk State University, Novosibirsk, Russia
{medvedev,kireev}@ssd.sscc.ru, ulia.trubiciina@gmail.com

Abstract. The paper discusses the implementation of cellular automata on supercomputers. It outlines the requirements for the software: ease of program construction and usability, ability to handle a wide range of transition functions, compatibility with various platforms, and ability to scale the size of cellular arrays with efficient use of computational resources. A review of software tools suitable for implementing cellular automata was conducted. One of these tools, a library of cellular automata topologies (CATlib), has been extended to implement synchronous cellular automata in parallel on multicomputers. The paper presents performance evaluation results emphasizing the high efficiency of the parallel implementation.

Keywords: Cellular automata · Parallel implementation · Software library · Supercomputer software · Simulation · Domain decomposition

1 Introduction

Cellular automaton (CA) is a mathematical model consisting of *cells* with discrete *states* combined into a regular spatial structure with local interactions, called *a cellular array*. The configuration of links between the cells is called *a topology*. The collective state of the cellular array changes iteratively. The state of each individual cell changes according to some discrete function called *a transition function*. Applying the transition function to the cells in a specific order and for a specific number of times is called *an iteration*.

CA was proposed back in the 1940s as a model of self-replicating systems [1], later they were used to study various physical, biological and chemical processes [2]. At the moment, there are many tools that simplify the development and study of CA models. However, most of them cannot provide the necessary

This work was carried out under the state contract with ICMMG SB RAS 0251-2022-0005.

V. Malyshkin (Ed.): PaCT 2023, LNCS 14098, pp. 93–105, 2023.
https://doi.org/10.1007/978-3-031-41673-6_8

functionality for a thorough study of large-scale processes, do not support large cellular arrays, and are mostly oriented to visualization instead of high computing performance. As a result, there is a need for a tool for developing software implementations of CA that meets the following criteria:

1. Simplicity of constructing CA models and convenience of their use.
2. Extensibility, i.e. supporting different types of CA with various transition functions.
3. Cross-platform, i.e. the ability of the system to run on different computer architectures and operating systems.
4. Scalability, i.e., with an increase in the size of the cellular array, the efficiency of using computing resources should remain acceptable.

In this paper, we will distinguish three functional categories of specialists involved in modeling: *a user, an application programmer* and *a system programmer*. The user runs a software product and solves the problem of engineering modeling without modifying any program code. An application programmer develops a simulation system, creates a program code for this system using software tools provided to him by a system programmer, which are designed to greatly facilitate his work. And finally, the system programmer develops these software tools for the application programmer. This is a generally accepted scheme.

Computer-aided engineering (CAE) systems are usually implemented in the form of three functional modules: *a preprocessor, a solver*, and *a postprocessor*. Using the preprocessor, the user sets the initial and boundary conditions of the problem, including geometry, parameters of materials and media. The solver finds a solution to the problem and almost does not interact with the user. In CA modeling methods, the role of solver is played by a simulator of the process under study. The postprocessor usually includes a viewer and various converters to present the simulation result to the user in a comprehensible way. If an application programmer follows this generally accepted architectural pattern when developing his CAE system, then a system programmer should provide a development tool as three sets of system routines: for the preprocessor, for the simulator and for the postprocessor.

In the context of high-performance computing, the pre- and postprocessor are of no interest to us, since their use in interactive mode usually does not cause difficulties. The simulator's system procedures apply the CA transition function, implemented by the application programmer in the form of a software component, to each cell of the cellular array repeatedly, a given number of iterations. They also implement the interaction of cells with each other without the participation of application software components. So, since the execution of simulator procedures turns out to be quite heavy due to the large number of repetitions, especially when processing large cellular arrays, the system programmer faces the task of repeatedly reducing the execution time of these procedures. Of course, the natural solution would be their parallel execution.

As mentioned above, CA operate iteratively. The order in which the cells of the array change their state during one iteration will be referred to as the *mode of operation* [3]. In *synchronous mode*, all cells of the array change their states simultaneously in accordance with the transition function. A common way to implement synchronous mode on a serial computer is to use a duplicate of the cellular array, into which the new cell states will be written during their traversal. In *asynchronous mode*, cells change their states at random moments in time, which is implemented by traversing them in random order.

Section 2 of the paper provides an overview of software tools that can be used to implement CA modeling systems. It is shown that none of the available tools satisfies all of the above criteria. The authors are developing the Cellular Automata Topologies Library (CATlib) [4], which is aimed at high performance and usability. Until recently, the library supported single-core execution only. In this paper, another step was taken in the development of the library, support for distributed memory systems for synchronous CA was implemented. Section 3 contains a brief description of the library and its programming interface. Section 4 describes the CATlib support for parallel distributed memory systems. Section 5 presents an example of CA, which is used for validation and performance evaluation of the library. Section 6 presents performance evaluation results. Section 7 concludes the paper.

2   Overview

Consider the existing software tools focused on the implementation of CA.

The ALT modeling system (Animating Language Tools) is designed for experiments with fine-grained parallel algorithms [5]. The system consists of a combination of text and graphical tools for visualizing fine-grained parallel computing. The processes are modeled using the parallel substitution algorithm [6]. To describe the constructions of this algorithm, ALT uses a special high-level language derived from the C language. The system also has various tools for textual and graphical editing of models and a number of tools for displaying the parameters of the computational process during modeling. The main advantages of using such a modeling system for research is the ability to monitor the calculation process and the ease of creating and editing various models. However, ALT does not support large cellular arrays and parallel processing. It was developed to run under the DOS operating system, and is now outdated and not supported by developers.

The system of simulation of algorithms with fine-grained parallelism WinALT is a continuation of the ideas of the ALT system [7,8]. The system has two versions: console and graphical. Modeling is performed using a special language designed to describe fine-grained parallel computing in the system, and also allows the use of functions written in C and C++ languages. WinALT has adopted all the advantages of the ALT system. In addition, it has the ability to select the operating mode of CA (synchronous, asynchronous, block-synchronous), create and edit modeling programs, and the maximum sizes of cellular arrays are slightly increased. The system does not provide for setting custom CA topologies, and the set of pre-implemented topologies is very scarce. WinALT was developed for Windows OS and may run on a Windows cluster. Currently, the system is not supported by developers.

Mirek's Cellebration (mCell) program was created to study existing and create new models of 1D and 2D CA [9]. The program supports 14 families of CA. It is possible to build CA with different types of neighborhood: Moore neighborhood, von Neumann neighborhood, Margolus neighborhood, and hexagonal neighborhood. For each family of CA, there are many built-in transition functions in the program: from well-known and well-studied to those developed by the author himself. Moreover, the user himself can add transition functions using external libraries in the ".dll" format. The presence of a graphical interface allows you to simply and quickly set the states of cells, as well as change these states during the evolution of the CA. mCell provides an opportunity to study CA with a cellular array size not exceeding 100000×2500 cells. There are also tools for collecting various statistics. mCell program may run only under Windows OS and does not support computing on a cluster. At the moment it is not supported by developers.

The Cafun (Cellular Automata Fun) application [10] is a tool for modeling complex systems, such as social groups, living organisms, natural processes, etc. The program was created to search for general laws of complex systems in order to better understand their development, structure and behavior. The practical result is the ability to give more accurate forecasts in economics, biology, physics and other areas where complexity plays a role. The concept of complex systems is based on three principles:

- Complex systems consist of many elements with individual properties and behavior.
- The properties of elements are the result of their local environment and individual history. Their behavior is determined by a limited and locally available amount of information without any centralization.
- The interaction between the elements occurs simultaneously. There is no pre-established sequence in which they occur.

The program uses the original notation of the concept of CA and an object-oriented approach to describing the laws of the CA. Its operation requires the Java Runtime Environment. The program is not intended to be used in distributed computing systems and currently is not supported by developers.

The Golly application is designed to study the behavior of various 1D, 2D and 3D CA [11]. Despite the fact that the developers position the application as an implementation of the Conway's Game of Life [12], Golly is a pretty powerful tool for building other models. In the application, it is possible to implement various classes of CA both using built-in software modules and using Python and Lua scripts. The application allows you to set different cellular array topologies. Efficient use of memory ensures work on cellular spaces of almost unlimited size, provided that most cells will still be empty, because only those cells that are not empty at the moment are processed. Golly can run on operating systems such as iOS, Android, Windows, Mac and Linux. As of August 2022, the system was supported by developers. The application does not allow calculations to be performed on the cluster.

The Tiled CA program was developed for the purpose of modeling by means of 1D and 2D CA [13]. The program allows you to split the modeling area in various ways, thus setting the neighborhood of each cell, but does not allow you to set a user-defined topology. The construction of CA models is carried out using a graphical editor, which allows you to quickly and easily edit the shape and states of cells. The program has few possibilities for constructing new transition functions, since Tiled CA supports only the variations of the Game of Life [12]. The program allows the user to set the size of the field, but they cannot exceed a certain fixed value. Tools for expanding the possibilities of building CA are not provided. Tiled CF only works on Windows, and it cannot be run on a cluster. Currently, the program is not supported by developers.

Wolfram Mathematica software has built-in libraries for a number of areas of technical computing, including for creating CA with some transition functions pre-defined in the system [14,15]. The software allows to implement 1D, 2D and 3D CA and supports a large number of transition functions for them. To implement a CA, it is required to choose one of the proposed transition functions, but at the same time Wolfram Mathematica provides opportunities for its modification. Wolfram Mathematica works under Windows, Linux, macOS, Android operating systems, and also has a web version. Upon completion, the program displays the final cellular array, but does not allow you to monitor its evolution.

The CelLab web application is designed to study various processes using 2D CA [16]. The application provides an opportunity to create CA by writing programs in Java and JavaScript, as well as to display the process of their evolution. The program allows you to set the transition function, the colors of cells depending on their state, control the display process and modify the transition function. The developers also provide a CelLab Development Kit archive, a wide range of ready-to-run transition functions that simulate various physical, chemical and biological processes. However, there is no way to run this application on a cluster and collect statistics.

The MATLAB application software package provides a wide range of possibilities for working with matrices. It also allows you to perform graphical multidimensional modeling of various systems. And although it is not a specialized tool for the implementation of CA, we mention it here, because attempts have been made to use it for the study of CA models [17,18]. The topology of the cellular array must be set by the user himself. The complexities of programming nontrivial topologies are fully present. In addition, the performance of the resulting code is much worse than that of implementations made manually in common programming languages.

As can be seen, the considered software tools do not solve the problem of high-performance implementation of large-scale CA. On the other hand, there are software tools for developing parallel programs that make it possible to develop large-scale CA models and use parallel computing systems for computations. Most of them require low-level system programming, which contradicts the first criterion. The easiest to use are shared-memory or PGAS-like tools, for example,

Coarray Fortran [19], DVM [20], UPC [21]. In addition, there are a number of tools that use special high-level programming models and relieve the user from the problems of system parallel programming [22,23]. A common disadvantage of all these parallel programming systems is that they do not take into account the specifics of CA algorithms and require writing additional code common to many implementations of CA. This is especially evident when implementing CA modes of operation other than synchronous.

3   The Cellular Automata Topologies Library

The Cellular Automata Topologies Library (CATlib) [4] is a set of system routines written in the C language that can be used by an application programmer to implement a modeling system (CAE) embodying the desired CA as a solver. Following the generally accepted architectural pattern described in the Introduction, the CATlib contains three subsets of system procedures: for implementing a preprocessor, for implementing a simulator, and for implementing a postprocessor.

The preprocessor converts the initial conditions in a physical formulation into cell states, and also initializes the service structure containing the dimensions of the cellular array and its topology, the size of memory required to store the cell state, the operating mode of the CA, etc. The following system procedures are available to the application programmer to implement the preprocessor.

CAT_InitPreprocessor – takes information about topology, model type, size of additional information, cell size in bytes, number of cells per meter and cellular array sizes; initializes the preprocessor environment in computer memory.

CAT_PutCell – takes the new state of the cell and its indexes in the cell array; writes the received state to the memory location corresponding to this cell.

CAT_FinalizePreprocessor – takes the name of the output file; saves the parameters of the CA set by the user and the state of the cellular array initialized by the user to a file in a special format defined by the library, frees the allocated memory.

The simulator iteratively applies the transition function of the CA to all cells of the array, taking into account a given operating mode, while ensuring the interaction of neighboring cells with each other in accordance with a given topology. The following system procedures are available to the application programmer to implement the simulator.

CAT_InitSimulator – takes the name of the input file containing the parameters of the CA and the state of the cellular array; initializes the control structure and the cellular array with data read from this file.

CAT_Iterate – takes a pointer to the procedure in which the application programmer implemented the transition function of the CA; applies the obtained function to the cells of the array, taking into account the operating mode that was set at initialization, performing one iteration of the CA.

`CAT_FileSave` – takes the name of the output file, saves the current state of the cellular array to the file.

`CAT_FinalizeSimulator` – takes the name of the output file; saves the resulting state of the cellular array, frees the allocated memory.

The postprocessor converts the states of the cells obtained as a result of the operation of the CA into a format that can be read and used by the user for further research. The following system procedures are available to the application programmer to implement the postprocessor.

`CAT_InitPostprocessor` – takes the name of the input file containing the parameters of the CA and the state of the cellular array; initializes the control structure and the cellular array with data read from this file.

`CAT_GetCell` – accepts cell indexes in the cellular array; returns its state.

`CAT_FinalizePostprocessor` – saves simulation results in formats understandable to the user, frees up allocated memory.

The implementation of the simulator for the Game of Life CA using the CATlib is shown in Listing 1.1.

Listing 1.1. Implementation of the simulator for the Game of Life CA

```
1  #include <stdio.h>
2  #include "catlib.h"
3
4  const int neighborsNumber = 8;
5  const int iterationsNumber = 100;
6
7  void gameOfLife(void *n){
8      int *cell = n;
9      int sum = 0;
10     for (int i = 0; i <= neighborsNumber; i++)
11         sum += cell[i];
12     cell[0] = (cell[0] | sum) == 3;
13  }
14
15  void main(int argc, char *argv[]){
16     CAT_InitSimulator("inputFileName", CAT_SYNC);
17     for (int i = 0; i < iterationsNumber; i++)
18         CAT_Iterate(gameOfLife);
19     CAT_FinalizeSimulator("outputFileName");
20  }
```

As can be seen from the Listing 1.1, to create a fully functional simulator, it is enough for an application programmer to implement only the transition function of his/her CA, in this case `gameOfLife()`. The library procedure `CAT_InitSimulator()` takes care of memory allocation for the cellular array and loads the cellular array and all necessary data from the file ''inputFileName'', including the selected topology, prepared by the preprocessor. The `CAT_Iterate()` procedure searches for the values of neighboring cells and applies the `gameOfLife()` transition function to each cell in synchronous mode. This procedure extracts the states of the current cell and its neighbors from the cell array and passes them to the transition function as a one-dimensional array `void *n`, and then returns the resulting state to the main array, hiding from the application programmer all the interactions of cells with each other. The `CAT_FinalizeSimulator()` procedure saves the result of the simulator in the form of a new cell array to the "`outputFileName`" file. This file has the same format as "`inputFileName`" and can be used both for postprocessing and for continuing the simulation by restarting the simulator.

The CATlib development plan is shown in Table 1.

Table 1. The CATlib development plan

Mode of operation	Computer architecture			
	Single-core CPU	Shared memory parallel	Distributed memory parallel	GPU
Synchronous	done	done	the present paper	planned
Asynchronous	done	in development	in development	planned

The library provides the application programmer with the means to implement synchronous and asynchronous CA for different computer architectures. It is available online[1]. Table 1 shows the development status for various library components. The purpose of this work is to develop support for synchronous CA on distributed computing systems.

4   Parallel Implementation

The algorithms and system procedures of the CATlib presented in this paper are designed to create parallel software implementations of synchronous cellular automata for distributed memory systems. They support the development of the simulator as a module of applied CAE systems. The preprocessor and postprocessor do not need parallel implementation as much as the simulator.

The 1D decomposition of the cellular array was chosen as the method of parallelization, since the difference in efficiency compared to the two-dimensional decomposition is small due to the high degree of inherent parallelism of CA [24]. Parallelization was performed using the MPI software interface. When running in parallel, the library automatically performs the following actions.

---

[1] https://gitlab.ssd.sscc.ru/medvedev/catlib

1. At the simulator initialization:
   (a) the master process reads the cellular array from the file in approximately equal parts according to the number of processes;
   (b) distribution of these parts among processes;
   (c) broadcasting the common task parameters read from the file to all processes.
2. During the execution of each iteration:
   (a) boundary exchange - each process transfers the updated states of boundary cells to neighboring processes;
   (b) application of the transition function to each cell of their own part of the cellular array.
3. On selected iterations and at the simulator finalization:
   (a) the master process collects parts of the resulting cellular array from all processes;
   (b) saving the assembled cellular array to a file;
   (c) saving common task parameters to a file.

From the application programmer's point of view, nothing changes when software components that support parallelism are added to the library. All program interfaces with the library remain unchanged. The only new requirement is that the MPI library is installed on the system.

5   Domino Cellular Automaton

We decided to test a new parallel version of the CATlib on a 2D CA forming a domino pattern (Domino CA in the following) taken from the literature [25–27]. We found this CA suitable for our purposes because its transition function has moderate complexity and its neighborhood is extended to a distance of two. Domino CA is an asynchronous 2D CA with a cellular array forming a square lattice with periodic boundaries. The cell's state is $\in \{$"0", "1", "#"$\}$. The neighborhood of a cell includes 24 cells forming a 5×5 square with a considered cell in the center. The transition rule is applied to a cell and, depending on its state and the states of its neighbors, can change its state. Cells with the state "#" are considered as "0" cells when checked, but they do not change their state during evolution. The purpose of the CA is to cover the cellular array with domino tiles of size 3×4 and 4×3, which are adjacent pairs of "1" cells (kernel) surrounded by "0" or "#" cells (hull). Hulls of different domino tiles are allowed to intersect, but kernels are not. An example of covering an area of size 10×10, surrounded by cells "#", is shown in Fig. 1.

The rule we have chosen to implement (combination of Basic rule and Rule option 2 in [26]) tries to maximize the number of domino tiles in the cellular array. It applies a predefined set of 5×5 templates to the neighborhood of the selected cell and counts the number of hits (h). Then, a new state is calculated depending on the value of h according to the formula (1).

$$state_{new} = \begin{cases} random \in \{0,1\}, \, if \, h = 0, \, applied \, with \, probability \, \pi_0, \\ random \in \{0,1\}, \, if \, h = 1, \, applied \, with \, probability \, \pi_1, \\ 0, \, if \, h = 2 \, or \, h = 3 \, or \, h = 4, \\ 1, \, if \, h = 100, \\ state_{old}, \, otherwise. \end{cases} \quad (1)$$

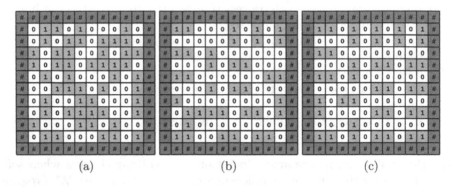

(a)   (b)   (c)

Fig. 1. Example of domino tiles covering an area of size 10×10 as a result of the Domino CA operation: iteration 0 (a), iteration 372 (b), iteration 1715 (c)

Each template contains the correct domino tile attached by one of its cells to the center of the template (horizontal and vertical tiles with 12 cells each give 24 templates). When applying a template to a cell and its neighborhood, only the non-central cells of the template that are part of the tile are checked. If all the states of the template cells being checked match the states of the corresponding neighboring cells, then the template is considered as matched and the counter h is incremented. The special case is when the template with "1" in the center is matched, in which the counter h is set to 100.

Parameters π_0 and π_1 (see formula 1) define the degree of random noise in case the value of h is small. We have found that the best values of these parameters, giving on average the fastest convergence to the maximum number of domino tiles, are $\pi_0 \in [0.4; 1.0]$, $\pi_1 \approx 0.005$ [28].

The CATlib allows one to create a parallel implementation of synchronous CA only. Therefore, to test the parallel version of the library, the mode of Domino CA operation was changed to synchronous. This transition is justified, because in synchronous mode, Domino CA can achieve maximum coverage of the cellular array with domino tiles in a number of iterations comparable to when asynchronous Domino CA is used [28]. The optimal values of the parameters π_0 and π_1 are the same in both modes. The noticed difference is that for synchronous Domino CA, the range of acceptable values of the π_1 parameter is narrower than for asynchronous Domino CA.

6 Performance Evaluation

To verify that the library allows for implementing complex cellular automata, and to evaluate the parallelization efficiency achieved with the library, we implemented the synchronous Domino CA described in the previous section. The following parameters were used: the size of the cellular array 10000 × 100 cells, the number of iterations is 1000. Tests were performed on the MVS-10P cluster of JSCC RAS [29]. The results are shown in Fig. 2.

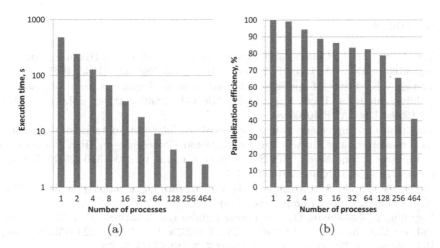

Fig. 2. Execution time (a) and parallelization efficiency (b) of synchronous Domino CA obtained with the CATlib

Figure 2a shows that the execution time in the case of one process was 483 s, and in the case of 256 processes it was 3 s. The efficiency of parallel implementation (Fig. 2b) was more than 65%, except for the last case of 464 processes, when we occupied the entire cluster. When using a single cluster node (16 cores), the parallelization efficiency is more than 85%.

The maximum CA size that we managed to run on the cluster was 60000 × 60000. Larger cellular arrays did not fit into the cluster's memory. The time it took to reach 95% of maximal domino coverage was about 15 h on 256 processes.

7 Conclusion

Appropriate software tools are required to develop large scale CA models. The criteria to be met by such tools are formulated. The review found that none of the considered software met all of these criteria. The paper presents the Cellular Automata Topologies Library, designed to satisfy all the criteria upon completion of development. The next step of its development is presented, which is the support of distributed computing systems for the implementation of synchronous CA. The new version of the library was tested on the synchronous version of Domino CA. Parallelization efficiency of more than 65% was achieved. The maximum size of the cellular array that we managed to process in a reasonable time was 3.6 gigacells.

Further research is planned to improve the computational performance of the parallel component of the library; in particular, it is planned to introduce a dynamic balancing algorithm similar to the one described in [30], which will improve the efficiency of the parallel implementation for some classes of cellular automata, such as lattice gas [24].

References

1. Von Neumann, J.: General and logical Theory of Automata, Hixon Symposium, reprinted in Taub, A.H. (ed.) Collected Works, vol. 288–328 (1948/1961)
2. Vanag, V.K.: Study of spatially extended dynamical systems using probabilistic cellular automata. Phys. Usp. **42**(5), 413–434 (1999). https://doi.org/10.1070/PU1999v042n05ABEH000558
3. Bandman O.: Implementation of large-scale cellular automata models on multi-core computers and clusters. In: International Conference on High Performance Computing & Simulation (HPCS), Helsinki, Finland, pp. 304–310 (2013). https://doi.org/10.1109/HPCSim.2013.6641431
4. Medvedev, Yu.G.: Architecture of the cellular automata topologies library. Bull. Nov. Comput. Center Comput. Sci. (46) (2022)
5. Pogudin, Y., Bandman, O.: Simulating cellular computations with ALT. A tutorial. In: Malyshkin, V. (ed.) PaCT 1997. LNCS, vol. 1277, pp. 424–435. Springer, Heidelberg (1997). https://doi.org/10.1007/3-540-63371-5_52
6. Achasova, S., Bandman, O., Markova, V., Piskunov, S.: Parallel Substitution Algorithm. Theory and Application. World Scientific Publ. (1994). https://doi.org/10.1142/2369. 232 p
7. Piskunov, S.: WinALT - a simulation system for computations with spatial parallelism. Bull. Nov. Comput. Center Comput. Sci. (6), 71–85 (1997)
8. Beletkov, D., Ostapkevich, M., Piskunov, S., Zhileev, I.: WinALT, a software tool for fine-grain algorithms and structures synthesis and simulation. In: Malyshkin, V. (ed.) PaCT 1999. LNCS, vol. 1662, pp. 491–496. Springer, Heidelberg (1999). https://doi.org/10.1007/3-540-48387-X_57
9. Mirek's Cellebration, 1-D and 2-D Cellular Automata viewer, explorer and editor. http://www.mirekw.com/ca/index.html. Accessed 1 May 2023
10. Homeyer, A.: A Brief Introduction To Cafun. https://cafun.de/information/a_brief_introduction_to_cafun/index.html. Accessed 1 May 2023
11. Golly Game of Life Home Page. https://golly.sourceforge.io. Accessed 1 May 2023
12. Gardner, M.: Mathematical Games - The fantastic combinations of John Conway's new solitaire game "life". Sci. Am. **223**(4), 120–123 (1970). https://doi.org/10.1038/scientificamerican1070-120
13. Tiled, C.A.: http://linuxenvy.com/bprentice/TiledCA/TiledCA.html. Accessed 1 May 2023
14. CellularAutomaton - Wolfram Language Documentation. https://reference.wolfram.com/language/ref/CellularAutomaton.html. Accessed 1 May 2023
15. Wolfram, S.: Cellular automata as models of complexity. Nature **311**, 419–424 (1984). https://doi.org/10.1038/311419a0

16. Cellular Automata Laboratory. https://www.fourmilab.ch/cellab/manual. Accessed 1 May 2023
17. Athanassopoulos, S., Kaklamanis, C., Kalfoutzos, G., Papaioannou, E.: Cellular automata: simulations using Matlab. In: Proceedings of the Sixth International Conference on Digital Society (ICDS), pp. 63–68 (2012)
18. Duarte Duarte, J.B., Talero Sarmiento, L.H., Sierra Juárez, K.J.: Evaluation of the effect of investor psychology on an artificial stock market through its degree of efficiency. Contaduríay Administración **62**(4), 1361–1376 (2017). https://doi.org/10.1016/j.cya.2017.06.014
19. Chivers, I., Sleightholme, J.: Coarray Fortran. In: Introduction to Programming with Fortran, pp. 501–512. Springer, Cham (2015). https://doi.org/10.1007/978-3-319-17701-4_32
20. Bakhtin, V.A., Krukov, V.A.: DVM-approach to the automation of the development of parallel programs for clusters. Program. Comput. Softw. **45**, 121–132 (2019). https://doi.org/10.1134/S0361768819030034
21. Carlson, W., Draper, J., Culler, D., et al.: Introduction to UPC and Language Specification. CCS-TR-99-157, IDA Center for Computing Sciences (1999)
22. Slaughter, E., Lee, W., Treichler, S., Bauer, M., Aiken, A.: Regent: a high-productivity programming language for HPC with logical regions. In: SC 2015: Proceedings of the International Conference for High Performance Computing, Networking, Storage and Analysis, Austin, TX, USA, pp. 1–12 (2015). https://doi.org/10.1145/2807591.2807629
23. Akhmed-Zaki, D., Lebedev, D., Malyshkin, V., Perepelkin, V.: Automated construction of high performance distributed programs in LuNA system. In: Malyshkin, V. (ed.) PaCT 2019. LNCS, vol. 11657, pp. 3–9. Springer, Cham (2019). https://doi.org/10.1007/978-3-030-25636-4_1
24. Medvedev, Yu.G.: Lattice gas cellular automata for a flow simulation and their parallel implementation. In: Tarkov, M.S. (ed.) Parallel Programming: Practical Aspects, Models and Current Limitations. Series: Mathematics Research Developments, pp. 143–158. Nova Science Publishers, Inc., Hauppauge, New York (2014)
25. Hoffmann, R., Désérable, D., Seredyński, F.: A probabilistic cellular automata rule forming domino patterns. In: Malyshkin, V. (ed.) PaCT 2019. LNCS, vol. 11657, pp. 334–344. Springer, Cham (2019). https://doi.org/10.1007/978-3-030-25636-4_26
26. Hoffmann, R., Désérable, D., Seredyński, F.: A cellular automata rule placing a maximal number of dominoes in the square and diamond. J. Supercomput. **77**, 9069–9087 (2021). https://doi.org/10.1007/s11227-020-03549-8
27. Hoffmann, R., Désérable, D., Seredyński, F.: Minimal covering of the space by domino tiles. In: Malyshkin, V. (ed.) PaCT 2021. LNCS, vol. 12942, pp. 453–465. Springer, Cham (2021). https://doi.org/10.1007/978-3-030-86359-3_35
28. Kireev, S., Trubitsyna, Yu.: Software implementation of asynchronous and synchronous cellular automata with maximum domino tiles coverage. Bull. Nov. Comput. Center Comput. Sci. (46) (2022)
29. Savin, G.I., Shabanov, B.M., Telegin, P.N., et al.: Joint supercomputer center of the Russian Academy of Sciences: present and future. Lobachevskii J. Math. **40**, 1853–1862 (2019). https://doi.org/10.1134/S1995080219110271
30. Medvedev, Y.: Dynamic load balancing for lattice gas simulations on a cluster. In: Malyshkin, V. (ed.) PaCT 2011. LNCS, vol. 6873, pp. 175–181. Springer, Heidelberg (2011). https://doi.org/10.1007/978-3-642-23178-0_15

Algorithms

Parallel-Batched Interpolation Search Tree

Vitaly Aksenov[1], Ilya Kokorin[1,2(✉)], and Alena Martsenyuk[1,2]

[1] ITMO University, Saint Petersburg, Russia
[2] vk.com, Saint Petersburg, Russia
kokorin.ilya.1998@gmail.com

Abstract. A sorted set (or map) is one of the most used data types in computer science. In addition to standard set operations, like `Insert`, `Remove`, and `Contains`, it can provide set-set operations such as `Union`, `Intersection`, and `Difference`. Each of these set-set operations is equivalent to some batched operation: the data structure should be able to execute `Insert`, `Remove`, and `Contains` on a batch of keys. It is obvious that we want these "large" operations to be parallelized. These sets are usually implemented with the trees of logarithmic height, such as 2–3 trees, treaps, AVL trees, red-black trees, etc. Until now, little attention was devoted to parallelizing data structures that work asymptotically better under several restrictions on the stored data. In this work, we parallelize Interpolation Search Tree which is expected to serve requests from a *smooth* distribution in doubly-logarithmic time. Our data structure of size n performs a batch of m operations in $O(m \log \log n)$ work and poly-log span.

Keywords: Parallel Programming · Data Structures · Parallel-Batched Data Structures

1 Introduction

A *Sorted set* is one of the most ubiquitous *Abstract Data Types* in Computer Science, supporting `Insert`, `Remove`, and `Contains` operations among many others. The sorted set can be implemented using different data structures: to name a few, skip-lists [21], red-black trees [11], splay trees [22], or B-trees [9, 10].

Since nowadays most of the processors have multiple cores, we are interested in parallelizing these data structures. There are two ways to do that: write a concurrent version of a data structure or allow one to execute a batch of operations in parallel. The first approach is typically very hard to implement correctly and efficiently due to problems with synchronization. Thus, in this work we are interested in the second approach: *parallel-batched data structures*.

Several parallel-batched data structures implementing a sorted set are presented: for example, 2–3 trees [18], red-black trees [17], treaps [6], (a, b) trees [2], AVL-trees [15], and generic joinable binary search trees [5, 23].

© The Author(s), under exclusive license to Springer Nature Switzerland AG 2023
V. Malyshkin (Ed.): PaCT 2023, LNCS 14098, pp. 109–125, 2023.
https://doi.org/10.1007/978-3-031-41673-6_9

Although many parallel-batched trees were presented, we definitely lack implementations that can execute separate queries in $o(\log n)$ time under some assumptions. However, there exist at least one sequential data structures with this property — *Interpolation Search Tree*, or *IST*.

Despite the fact that concurrent IST is already presented [8,20] we still lack its parallel-batched version: it differs much from the concurrent version since it allows many processes to execute scalar requests simultaneously, while we use the multiprocessing to parallelize large non-scalar requests.

The work is structured as follows: in Sect. 2 we describe the important preliminaries; in Sect. 3 we present the original Interpolation Search Tree; in Sects. 4, 5, and 6 we present the parallel-batched contains, insert and remove algorithms; in Sect. 7 we present a parallelizable method to keep the IST balanced; in Sect. 8 we present a theoretical analysis; in Sect. 9 we discuss the implementation and present experimental results; we conclude in Sect. 10. The full version of the paper appears at [3].

2   Preliminaries

2.1   Parallel-Batched Data Structures

Definition 1. Consider a data structure D storing a set of keys and an operation Op. If Op involves only one key (e.g., it checks whether a single key exists in the set, or inserts a single key into the set) it is called a *scalar operation*. Otherwise, i.e., if Op involves multiple keys, it is called a *batched operation*.

A data structure D that supports at least one *batched operation* is called a *batched data structure*.

We want the following batched operations from a sorted set:

- `Set.ContainsBatched(keys[])` — the operation takes an array of keys of size m and returns an array `Result` of size m. For each i, `Result[i]` is `true` if `keys[i]` exist in the set, and `false` otherwise.
- `Set.InsertBatched(keys[])` — the operation takes an array of keys of size m. If `keys[i]` does not exist in the set, the operation adds it to the set.
- `Set.RemoveBatched(keys[])` — the operation takes an array of keys of size m. If `keys[i]` exists in the set, the operation removes it from the set.

Note, that: 1) `InsertBatched` calculates the union of two sets; 2) `RemoveBatched` calculates the difference of two sets; and 3) `ContainsBatched` calculates the intersection of two sets.

We can employ parallel programming techniques (e.g., *fork-join parallelism* [7,14]) to execute batched operations faster.

Definition 2. *A batched data structure D that uses parallel programming to speed up batched operation execution is called a* parallel-batched data structure.

2.2   Time Complexity Model

In our work, we assume the standard *work-span* complexity model [1] for *fork-join* computations. We model each computation as a *directed acyclic graph*, where nodes represent operations and edges represent dependencies between them. This graph has exactly one *source node* (i.e., the start of the execution with zero incoming edges) and exactly one *sink node* (i.e., the end of the execution with zero outcoming edges). Some operations have two outcoming edges — they spawn two parallel tasks and are called *fork operations*. Some operations have two incoming edges — they wait for two corresponding parallel tasks to complete and are called *join operations*.

Considering the execution graph of the algorithm, our target complexities are: 1) *work* denotes the number of nodes in the graph, i.e., the total number of operations executed; 2) *span* denotes the number of nodes on the longest path from *source* to *sink*, i.e., the length of the critical path in the graph.

2.3   Standard Parallel Primitives

In this work, we use several standard parallel primitives. Now, we give their descriptions. Their implementations are provided, for example, in [12].

Parallel loop. It executes a loop body for n index values (from 0 to n-1, inclusive) in parallel. This operation costs $O(n)$ work and $O(\log n)$ span given that the body has time complexity $O(1)$.

Scan. Result := Scan(Arr) calculates *exclusive* prefix sums of array Arr such that $Result[i] = \sum_{j=0}^{i-1} Arr[j]$. Scan has $O(n)$ work and $O(\log n)$ span.

Filter. Filter(Arr, predicate) returns an array, consisting of elements of the given array Arr satisfying predicate keeping the order. Filter has $O(n)$ work and $O(\log n)$ span given that predicate has time complexity $O(1)$.

Merge. Merge(A, B) merges two sorted arrays A and B keeping the result sorted. It has $O(|A| + |B|)$ work and $O(\log^2(|A| + |B|))$ span.

Difference. Difference(A, B) takes two sorted arrays A and B and returns all elements from A that are not present in B, in sorted order. It takes $O(|A|+|B|)$ work and $O(\log^2(|A| + |B|))$ span.

Rank.   Given that A is a sorted array and x is a value, we denote ElemRank(A, x) $= |\{e \in A | e \leq x\}|$ as the number of elements in A that are less than or equal to x. Given that A and B are sorted arrays, we denote Rank(A, B) = $[r_0, r_1, ..., r_{|B|-1}]$, where r_i = ElemRank(A, B[i]). Rank operation can be computed in $O(|A| + |B|)$ work and $O(\log^2(|A| + |B|))$ span.

3   Interpolation Search Tree

3.1   Interpolation Search Tree Definition

Interpolation Search Tree (IST) is a multiway internal search tree proposed in [16]. IST for a set of keys $x_0 < x_1 < ... < x_{n-1}$ can be either *leaf* or *non-leaf*.

Definition 3. Leaf *IST with a set of keys* $x_0 < x_1 < \ldots < x_{n-1}$ *consists of array* `Rep` *with* $Rep[i] = x_i$, *i.e., it keeps all the keys in this sorted array.*

Definition 4. Non-leaf *IST with a set of keys* $x_0 < x_1 < \ldots < x_{n-1}$ *consists of two parts (Fig. 1 and 2):*

- *An array* `Rep` *storing an ordered subset of keys* $x_{i_0}, x_{i_1}, \ldots x_{i_{k-1}}$.
- *Child ISTs* $C_0, C_1 \ldots C_k$: *1)* C_0 *is an IST storing a subset of keys* $x_0, x_1 \ldots x_{i_0-1}$; *2) for* $1 \leq j \leq k - 1$, C_j *is an IST storing a subset of keys* $x_{i_{j-1}+1}, \ldots x_{i_j-1}$; *and 3)* C_k *is an IST storing a subset of keys* $x_{i_{k-1}+1}, \ldots x_{n-1}$;

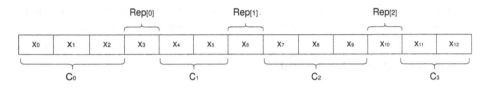

Fig. 1. Example of a non-leaf IST. $Rep[0] = x_3, Rep[1] = x_6, Rep[2] = x_{10}$. C_0 stores keys $x_0 \ldots x_2$, C_2 stores keys $x_4 \ldots x_5$, C_3 stores keys $x_7 \ldots x_9$, C_4 stores keys $x_{11} \ldots x_{12}$.

Any *non-leaf* IST has the following properties: 1) all keys less than $Rep[0]$ are located in C_0; 2) all keys in between $Rep[j-1]$ and $Rep[j]$ are located in C_j, and, finally, 3) all keys greater than $Rep[k-1]$ are located in C_k.

3.2 Interpolation Search and the Lightweight Index

We can optimize operations on ISTs with numeric keys, by leveraging the *interpolation search technique* [16,19, 24]. Each node of an IST has an index that can point to some place in the `Rep` array close to the position of the key being searched. This approach is named *interpolation search*. The structure of a non-leaf IST with an index is shown in Fig. 3.

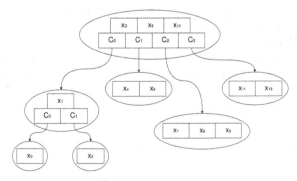

Fig. 2. Example of an IST built on array in Fig. 1.

In the original IST, the index uses an array ID of size $m \in \Theta(n^\varepsilon)$ with some $\varepsilon \in [\frac{1}{2}; 1)$. ID[i] = j iff $Rep[j] < a + i \cdot \frac{b-a}{m} \leq Rep[j+1]$ where a and b are the lower and upper bounds on the values. In [16], ID$[\lfloor \frac{x-a}{b-a} \cdot m \rfloor]$ is the approximate position of x in Rep.

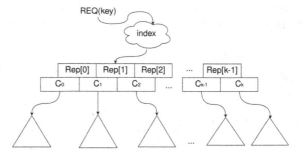

Fig. 3. Non-leaf IST contains: (1) Rep array; (2) an array of pointers to child ISTs C; (3) an index, allowing for fast lookups of keys in the Rep array.

After finding the approximate location of x in Rep, we can find its exact location by using the linear search, as described in [16]. Let us denote i := ID$[\lfloor \frac{x-a}{b-a} \cdot m \rfloor]$. If i points to the right place — we stop. Otherwise, we go in the proper direction: to the right of i (Fig. 4a) or to the left of i (Fig. 4b).

Note, we can use more complex techniques instead of the linear search, e.g., exponential search [4]. However, they are often unnecessary, since the index usually provides an approximation good enough to finish the search only in a couple of operations. Also, we can use other index structures, e.g., a machine learning model [13].

3.3 Search in IST

Suppose we want to find a key in IST. The search algorithm is iterative: on each iteration we look for the key in a subtree of a node v. To look for the key in the whole IST we begin the algorithm by setting v := IST.Root.

To find key in v, we do the following (k is the length of v.Rep):

1. If v is empty, we conclude that key is not there;
2. If key is found in v.Rep array, then, we found the key;
3. If key < v.Rep[0], the key can be found only in v.C[0] subtree. Thus, we set v ← v.C[0] and continue the search;
4. If key > v.Rep[k - 1], the key can be found only in v.C[k] subtree. Thus, we set v ← v.C[k] and continue the search;
5. Otherwise, we find j such that v.Rep[j - 1] < key < v.Rep[j]. In this case key can be found only in v.C[j]. Thus, we set v ← v.C[j] and continue our search in the j-th child.

(a) Searching for the key on the right to the approximate position.

(b) Searching for the key on the left to the approximate position.

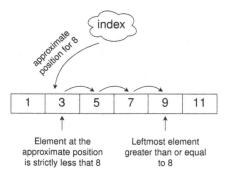 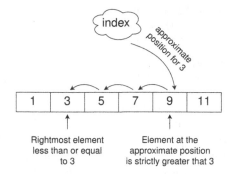

Fig. 4. Determining the exact location of the key given the approximate location

3.4   Executing Update Operations and Maintaining Balance

The algorithm for inserting a key into IST is very similar to the search algorithm above. To execute `Insert(key)` we do the following (Fig. 5):

1. Initialize `v := IST.Root`;
2. For the current node v, if **key** appears in `v.Rep` array, we finish the operation — the key already exists.
3. If v is a leaf and **key** does not exists in `v.Rep`, insert **key** into `v.Rep` keeping it sorted;
4. If v is an inner node and **key** does not exists in `v.Rep`, determine in which child the insertion should continue, set `v ← v.C[next_child_idx]` and go to step (2).

To remove a key from IST we introduce `Exists` array in each node that shows whether the corresponding key in `Rep` is in the set or not. Thus, we just need to mark an element as removed without physically deleting it. We have to take into account such marked keys during the contains and inserts. The removal algorithm is discussed in more detail in [16].

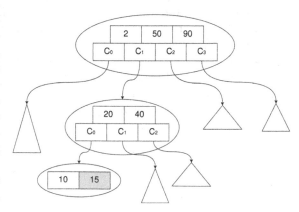

Fig. 5. Insert 15: proceed from the root to the second child and then to the first child.

The problem with these update algorithms is that all the new keys may be inserted to a single leaf, making the IST unbalanced. In order to keep the execution time low, we should keep the tree balanced.

Definition 5. *Suppose H is some small integer constant, e.g., 10. An IST T, storing keys $x_0 < x_1 < \ldots < x_{n-1}$, is said to be ideally balanced if either: 1) T is a leaf IST and $n \leq H$; 2) T is a non-leaf IST, $n > H$, and elements in Rep are equally spaced, $Rep[i] = x_{(i+1) \cdot \lfloor \frac{n}{k} \rfloor}$, and all child ISTs $\{C_i\}_{i=0}^{k}$ are ideally balanced.*

For non-leaf IST, we aim to have the size of Rep as $k = \lfloor \sqrt{n} \rfloor$. Consider an ideally balanced IST storing n keys (Fig. 6). As we require, the root of IST contains $\Theta(n^{\frac{1}{2}})$ keys in its Rep array; any node on the second level has Rep array of size $\Theta(n^{\frac{1}{4}})$; generally, any node on the i-th level has Rep array of size $\Theta(n^{\frac{1}{2^i}})$. Thus, an ideally balanced IST with n keys has a height of $O(\log \log n)$.

In order to keep IST balanced we maintain the number of modifications (both insertions and removals) applied to each subtree T. When the number of modifications to T exceeds the initial size of T multiplied by some constant C, we rebuild T from scratch making it ideally balanced. This rebuilding approach has a proper amortized bounds and is adopted from papers about IST [8,16,20].

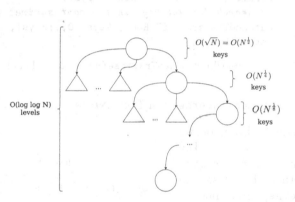

Fig. 6. Height of an ideal IST

3.5 Time and Space Complexity

Mehlhorn and Tsakalidis [16] define of a *smooth* probability distribution. For example, the uniform distribution is *smooth*. Suppose we are given μ that is *smooth*. From [16] we know that: 1) IST with n keys takes $O(n)$ space; 2) the expected amortized cost of μ-random insertion and random removal is $O(\log \log n)$; 3) the amortized insertion and removal cost is $O(\log n)$; 4) the expected search time on sets, generated by μ-random insertions and random removal, is $O(\log \log n)$; 5) the worst-case search time is $O(\log^2 n)$.

Therefore, IST can execute operations in $o(\log n)$ time under reasonable assumptions. As our goal, we want to design a parallel-batched version of the IST that processes operations asymptotically faster than known sorted set implementations (e.g., red-black trees).

4   Parallel-Batched Contains

In this section, we describe the implementation of ContainsBatched(keys[]) operation. We suppose that keys array is sorted. For simplicity, we assume that IST does not support removals. In Sect. 6, we explain how to fix it.

We implement ContainsBatched operation in the following way. At first, we introduce a function BatchedTraverse(node, keys[], left, right, result[]). The purpose of this function is to determine for each index left $\leq$ $i <$ right, whether keys[i] is stored in the node subtree. If so, set result[i] = true, otherwise, result[i] = false. Given the operation BatchedTraverse, we can implement ContainsBatched with almost zero effort (Listing 1.1):

Listing 1.1. Implementation of ContainsBatched on top of BatchedTraverse routine

```
fun ContainsBatched(keys[]):
    result[] := [array of size |keys|]
    // search for all keys in the root subtree (i.e., in the whole IST)
    BatchedTraverse(IST.Root, keys, 0, |keys|, result)
    return result
```

Now, we describe BatchedTraverse(node, keys[], left, right, result[]).

4.1   BatchedTraverse in a Leaf Node

If node is a leaf node, we determine for each key in keys[left..right) whether it exists in node.Rep. Since node is a leaf, keys cannot be found anywhere else in node subtree.

We may use Rank function to find the *rank* of each element of keys[left..right) in node.Rep and, thus, determine for each key

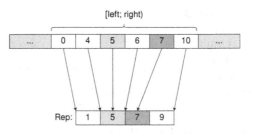

Fig. 7. Execution of BatchedTraverse in an IST leaf. Here Rank(node.Rep, keys[left..right)) = [1, 1, 2, 2, 3, 4].

whether it exists in node.Rep (Fig. 7, Listing 1.2). As presented in Sect. 2.3, ranks of all keys from subarray keys[left..right) may be computed in parallel in linear work and poly-logarithmic span.

Listing 1.2. Using `Rank` to find keys in a leaf node in parallel

```
rank := Rank(node.Rep, keys[left..right))
pfor i in left..right:
    r := rank[i - left]
    if r = 0 or node.Rep[r - 1] ≠ keys[i]:
        result[i] ← false
    else:
        result[i] ← true
```

4.2 `BatchedTraverse` in an Inner Node

Consider now the `BatchedTraverse` procedure on an inner node (Fig. 8).

We begin its execution with finding the position for each key from `keys[left..right)` in `node.Rep`. We may do it using `Rank` function as in Sect. 4.1. However, we can also use the interpolation search (described in Sect. 3.2): see Listing 1.3. Denote T as an interpolation search time in `node.Rep`. As stated in Sect. 3.5, T is expected to be $O(1)$. Thus, this algorithm can be executed in $O((right - left) \cdot T)$ work and polylog span in contrast to the algorithm based on the `Rank` function, that takes $O((right - left) + |node.Rep|)$ work.

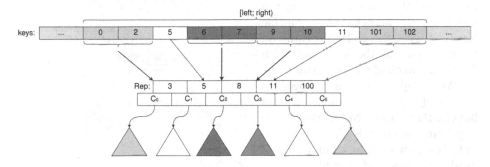

Fig. 8. Execution of `BatchedTraverse` in an inner node of an IST.

Listing 1.3. Using interpolation search to find keys in IST leaf node in parallel

```
pfor i in left..right:
    idx := interpolation_search(node.Rep, keys[i])
    result[i] ← node.Rep[idx] = keys[i]
```

Some keys of the input array (e.g., 5 and 11 in Fig. 8) are found in the `Rep` array. For such keys, we set `Result[i]` to `true`. After that, all other keys can be divided into three categories:

- Keys that are strictly less than `Rep[0]` (e.g., 0 and 2 in Fig. 8) lie in `C[0]` subtree. Therefore for such keys we should continue the traversal in `C[0]`;
- Keys that are strictly greater than `Rep[k - 1]` (e.g., 100 and 101 in Fig. 8) can only be found in `C[k]`. Therefore for such keys we continue the traversal in `C[k]`.

– Keys that lie between Rep[i] and Rep[i + 1] for some $i \in [0; k-2]$ (we can find such i for each key using the same search technique as described above). For example, 6 and 7 for i = 1 or 9 and 10 for i = 2 in Fig. 8. Such keys can only be found in C[i + 1]. Therefore for such keys we should continue the traversal in C[i + 1];

Note that some child nodes (e.g., C[1] and C[4] in Fig. 8) can not contain any key from keys[left..right) thus we do not continue the search in such nodes.

After determining in which child the search of each key should continue we proceed to searching for keys in children in parallel.

5 Parallel-Batched Insert

We now consider the implementation of the operation InsertBatched(keys[]). Again, we suppose that array keys[] is sorted. For simplicity, we consider InsertBatched implementation on an IST without removals. In Sect. 6 we explain how to fix it.

We begin the insertion procedure with filtering out keys already present in the set. We can do this using the described ContainsBatched routine together with the Filter primitive: we filter out all the keys for which ContainsBatched returns true.

We implement our procedure recursively in the same way as BatchedTraverse. Note, that each key being inserted is not present in IST, thus, for each key our traversal finishes in some leaf (Fig. 9).

After we finish the traversal — we need to insert subarray keys[left$_i$ ... right$_i$) into some leaf leaf$_i$. For example, in Fig. 9 we insert keys[0..2) (i.e., 0 and 3) to the leftmost leaf, while inserting keys[2..4) (i.e., 18 and 19) to the rightmost leaf.

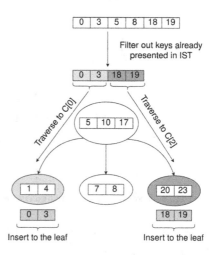

Fig. 9. Inserting a batch of keys in the IST

To finish the insertion, we just merge keys[left$_i$ ... right$_i$) with leaf$_i$.Rep and get the new Rep array. Now, each target leaf leaf$_i$ contains all the keys that should be inserted into it.

6 Parallel-Batched Remove

We now sketch the implementation of the operation RemoveBatched(keys[]). Again, we suppose that array keys is sorted.

We use the same approach as our previous algorithms. At first, we filter out the keys that do not exist in the tree. Then, we go recursively, find the keys in Rep arrays, and set the corresponding Exists cell to false.

Since now we have a logical removal, we should modify the implementations of ContainsBatched and InsertBatched.

During the execution of ContainsBatched when we encounter the key being searched in the Rep array of some node v (v.Rep[i] = key), we check v.Exists[i]: 1) if v.Exists[i] = true then key exists in the set; 2) otherwise, key does not exist in the set.

Now, we explain the updates to InsertBatched. As was stated in Sect. 5, we cannot encounter any of the key being inserted in the Rep array of any node of IST, since we filter out all the keys existing in IST. However, when keys can be logically removed this is not true anymore. Such keys have the corresponding entry in v.Exists array set to false, since the key does not logically exist in IST (Fig. 10a).

Suppose we are inserting key and we encounter it in some v.Rep. Thus, we can just set v.Exists[i] ← true (Fig. 10b). This way the insert operation "revives" a previously removed key.

(a) Keys 1, 10 and 23 are marked as removed

(b) Keys 10 and 23 are revived by a subsequent insert operation

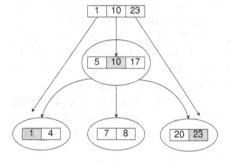 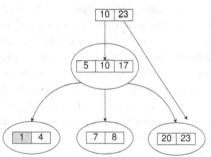

Fig. 10. Insertion of a key, that still exists in the IST physically, but is removed logically

7   Parallel Tree Rebuilding

7.1   Rebuilding Principle

As stated in Sect. 3.4, we employ the lazy subtree rebuilding approach to keep IST balanced. This algorithm is adopted from papers [8,16,20].

For each node of IST we maintain Mod_Cnt — the number of modifications (successful insertions and removals) applied to that node subtree. Moreover, each node stores Init_Subtree_Size — the number of keys in that node subtree when the node was created.

Suppose we execute an update operation Op in node v and Op increases v.Mod_Cnt by k (i.e., it either inserts k new keys or removes k existing keys).

If v.Mod_Cnt + k ≤ C · v.Init_Subtree_Size (where C is a predefined constant, e.g., 2) we increment v.Mod_Cnt by k and continue the execution of Op in an ordinary way. Otherwise, we rebuild the whole subtree of v.

The subtree rebuilding works in the following way. At first, we *flatten* the subtree into an array: we collect all non-removed keys from the subtree to array subtree_keys[] in ascending order. This operation is described in more detail in Sect. 7.2. If the operation, that triggered the rebuilding, was InsertBatched, we merge the keys, we are inserting, with the keys from the subtree_keys. Otherwise, (that operation is RemoveBatched) we remove the required keys from the subtree_keys via the Difference operation (see Sect. 2.3 for details). Finally, we build an ideal IST new_subtree, containing all entries from subtree_keys. This operation is described in more detail in Sect. 7.3.

7.2   Flattening an IST into an Array in Parallel

First of all, we need to know how many keys are located in each node subtree. We store this number in a Size variable in each node and maintain it the following way: 1) when creating new node v, set v.Size to the initial number of keys in its subtree; 2) when inserting m new keys to v's subtree, increment v.Size by m; 3) when removing m existing keys from v's subtree, decrement v.Size by m.

To flatten the whole subtree of **node** we allocate an array **keys** of size **node.Size** where we shall store all the keys from the subtree. We implement the flattening recursively, via the Flatten(v, keys[], left, right) procedure, which fills subarray keys[left..right) with all the keys from the subtree. To flatten the whole subtree of **node** into newly-allocated array **subtree_keys** of size **node.Size** we use Flatten(node, subtree_keys, 0, node.Size).

Note that non-leaf node v has 2k + 1 sources of keys: v.C[i] with v.C[i].Size keys and v.Rep[i]. C[i] is 2 · i-th key source and Rep[i] is 2 · i + 1-th key source. Note that for a leaf node all children just contain 0 keys.

Now for each key source we must find its position in the **keys** array. To do this we calculate array **sizes** of size 2k + 1. i-th source of keys stores its keys count in sizes[i]. After that we calculate positions := Scan(sizes) to find the prefix sums of sizes. After that positions[i] = $\sum_{j=0}^{i-1}$ sizes[j]. Consider now i-th key source. All prior key sources should fill positions[i] keys, thus, i-th key source should place its keys into the array starting from left + positions[i] position (Fig. 11). Therefore:

- v.C[i] places its keys in the **keys** array starting from left + positions[2· i] by running Flatten(v.C[i], left + positions[2 · i], left + positions[2 · i] + v.C[i].Size).
- If v.Exists[i] = false then v.Rep[i] should not be put in the **keys** array;
- Otherwise, v.Exists[i] = true and v.Rep[i] should be placed at keys[left + positions[2 · i + 1]] since v.Rep[i] is the (2 · i + 1)-th key source.

Each key source can be processed in parallel, since there are no data dependencies between them.

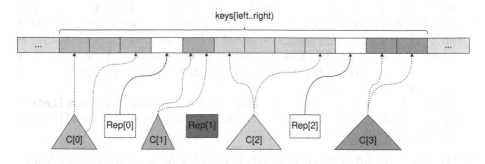

Fig. 11. Parallel flattening of an IST node

In Fig. 11, C[0] will place its keys in keys[left .. left + 3) subarray, Rep[0] will be placed in keys[left + 3], C[1] will place its keys in keys[left + 4 .. left + 5) subarray, Rep[1] will not be placed in keys array since its logically removed, C[2] will place its keys in keys[left + 5 .. left + 9) subarray, Rep[2] will be placed in keys[left + 9] and C [3] will place its keys in keys[left + 10 .. left + 12) subarray.

7.3   Building an Ideal IST in Parallel

Suppose we have a sorted array of keys and we want to build an *ideally balanced IST* (see Sect. 3.4) with these keys. We implement this procedure recursively via build_IST_subarray(keys[], left, right) procedure — it builds an ideal IST containing keys from the keys[left..right) subarray and returns the root of the newly-built subtree. Thus, to build a new subtree from array keys we just use new_root := build_IST_subarray(keys[], 0, |keys|).

If the size of the subarray (i.e., right - left) is less than a predefined constant H, we return a leaf node with all the keys from keys[left..right) in Rep array.

Otherwise (i.e., if right - left is big enough), we have to build non-leaf node. Let us denote m := right - left; k := $\lfloor\sqrt{m}\rfloor - 1$. As follows from Definition 5, Rep array should have size $\Theta(\sqrt{m})$ and its elements must be equally spaced keys of the initial array. Thus, we copy each k-th key (k-th, $2 \cdot k$-th, etc.) into array Rep. Note, our subarray begins at position left of the initial array, since we are building IST from the subarray keys[left..right). Thus, we copy keys[left + (i + 1) · k] into Rep[i]. All the copying can be done in parallel since there are no data dependencies. This way we obtain Rep array of size $\Theta(\sqrt{m})$ filled with equally-spaced keys of the initial subarray (Fig. 12).

Now we should build the children of the newly-created node (Fig. 12):

- Rep[0] = keys[left + k]. Thus, all keys less than keys[left + k] will be stored in C[0] subtree: C[0] ← build_IST_subarray(keys[], left, left + k);
- for 1 ≤ i ≤ k − 2, Rep[i - 1] = keys[left + i · k] and Rep[i] = keys[left + (i + 1) · k]. Thus, all keys x such that Rep[i-1] < x < Rep[i] should be stored in C[i] subtree. Since keys array is sorted, C[i] must be built from the subarray keys[left + i·k + 1..left + (i + 1) · k), thus, C[i] ← build_IST_subarray(keys[], left + i · k + 1, left + (i + 1) · k);
- Rep[k - 1] = keys[left + k^2]. Thus, all keys greater than keys[left + k^2] are stored in C[k] subtree: C[k] ← build_IST_subarray(keys[], left + k^2, right).

We can build all children in parallel, since there is no data dependencies between them.

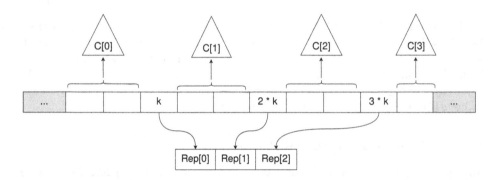

Fig. 12. Building children of a new node

To finish the construction of a node we need to calculate node.ID array described in Sect. 3.2. We can build it in the following way:

- Create an array bounds of size m + 1 such that bound[i] = keys[left] + i · (keys[right - 1] − keys[left])/L where $L = \Theta(n^\varepsilon)$ with some $\varepsilon \in [\frac{1}{2}; 1)$;
- Use Rank primitive to find the rank of each bounds[i] in the Rep array.

8 Theoretical Results

In this section, we present the theoretical bounds for our data structure. These bounds are quite trivial, so we just give intuition.

Theorem 1. *The flatten operation of an IST with n elements has $O(n)$ work and $O(\log^3 n)$ span. The building procedure of an ideal IST from an array of size n has $O(n)$ work and $O(\log n \cdot \log \log n)$ span. Thus, the rebuilding of IST with n elements costs $O(n)$ work and $O(\log^3 n)$ span.*

Proof (Sketch). While the work bounds are trivial, we are more interested in span bounds. From [16] we know that in the worst case, the height of IST with n keys does not exceed $O(\log^2 n)$. Thus, the flatten operation just goes recursively into $O(\log^2 n)$ levels and spends $O(\log n)$ span at each level. This gives $O(\log^3 n)$ span in total. The construction of an ideal IST has $O(\log \log n)$ recursive levels while each level can be executed in $O(\log n)$ time, i.e., copy the elements into Rep array. This gives us the result.

This brings us closer to our main complexity theorem.

Theorem 2. *The work of a batched operation on our parallel-batched IST has the same complexity as if we apply all m operations from this batch sequentially to the original IST of size n (from [16], the expected execution time is $O(m \log \log n)$). The total span of a batched operation is $O(\log^4 n)$.*

Proof (Sketch). The work bound is trivial — the only difference with the original IST is that we can rebuild the subtree in advance before applying some of the operations. Now, we get to the span bounds. From [16], we know that the height of IST with n keys does not exceed $O(\log^2 n)$. On each level, we spend: 1) at most $O(\log^2 n)$ span for merge and rank operations; or 2) we rebuild a subtree at that level and stop. The first part gives us $O(\log^4 n)$ span, while rebuilding takes just $O(\log^3 n)$ span. This leads us to the result of the total $O(\log^4 n)$ span.

9   Experiments

We have implemented the Parallel Batched IST in C++ using OpenCilk [7] as a framework for fork-join parallelism.

We tested our parallel-batched IST on three workloads. We initialize the tree with elements from the range $[-10^8; 10^8]$ with probability $1/2$. Thus, the expected size of the tree is 10^8. Then we call: 1) search for a batch of 10^7 keys, taken uniformly at random from the range; 2) insert a batch of random 10^7 keys, taken uniformly at random from the range; 3) remove a batch of random 10^7 keys, taken uniformly at random from the range.

The experimental results are shown in Fig. 13. The OX axis corresponds to the number of worker processes and the OY axis corresponds to the time required to execute the operation in milliseconds. Each point of the plot is obtained as an average of 10 runs. We run our code on an Intel Xeon Gold 6230 machine with 16 cores.

As shown in the results, we achieve good scalability. Indeed: 1) 14x scaling on ContainsBatched operation for 16 processes; 2) 11x scaling on InsertBatched operation for 16 processes; 3) 13x scaling on RemoveBatched operation for 16 processes.

We also compared our implementation in a sequential mode with std::set. std::set took 9257 ms to check the existence of 10^7 keys in a tree with 10^8 elements while our IST implementation took only 3561 ms. We achieve such speedup by using interpolation search as described in Sect. 3.2.

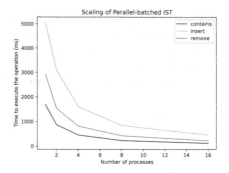

Fig. 13. Benchmark results for Parallel-batched Interpolation Search Tree

10   Conclusion

In this work, we presented the first parallel-batched version of the interpolation search tree that has an optimal work in comparison to the sequential implementation and has a polylogarithmic span. We implemented it and got very promising results. We believe that this work will encourage others to look into parallel-batched data structures based on something more complex than binary search trees.

References

1. Acar, U.A., Blelloch, G.E.: Algorithms: Parallel and sequential. https://www.umut-acar.org/algorithms-book 6 (2019)
2. Akhremtsev, Y., Sanders, P.: Fast parallel operations on search trees. In: 2016 IEEE 23rd International Conference on High Performance Computing (HiPC), pp. 291–300. IEEE (2016)
3. Aksenov, V., Kokorin, I., Martsenyuk, A.: Parallel-batched interpolation search tree. arXiv preprint (2023). arXiv:2306.13785
4. Bentley, J.L., Yao, A.C.C.: An almost optimal algorithm for unbounded searching. Inform. Process. Lett. **5**(3), 82–87(SLAC-PUB-1679) (1976)
5. Blelloch, G.E., Ferizovic, D., Sun, Y.: Just join for parallel ordered sets. In: Proceedings of the 28th ACM Symposium on Parallelism in Algorithms and Architectures, pp. 253–264 (2016)
6. Blelloch, G.E., Reid-Miller, M.: Fast set operations using treaps. In: Proceedings of the Tenth Annual ACM Symposium on Parallel Algorithms And Architectures, pp. 16–26 (1998)
7. Blumofe, R.D., Joerg, C.F., Kuszmaul, B.C., Leiserson, C.E., Randall, K.H., Zhou, Y.: Cilk: an efficient multithreaded runtime system. J. Parallel Distrib. Comput. **37**(1), 55–69 (1996)
8. Brown, T., Prokopec, A., Alistarh, D.: Non-blocking interpolation search trees with doubly-logarithmic running time. In: Proceedings of the 25th ACM SIGPLAN Symposium on Principles and Practice of Parallel Programming, pp. 276–291 (2020)

9. Comer, D.: Ubiquitous b-tree. ACM Comput. Surv. (CSUR) **11**(2), 121–137 (1979)
10. Graefe, G., et al.: Modern b-tree techniques. Foundations and Trends® in Databases **3**(4), 203–402 (2011)
11. Guibas, L.J., Sedgewick, R.: A dichromatic framework for balanced trees. In: 19th Annual Symposium on Foundations of Computer Science (sfcs 1978), pp. 8–21. IEEE (1978)
12. JáJá, J.: An introduction to parallel algorithms. Reading, MA: Addison-Wesley **10**, 133889 (1992)
13. Kraska, T., Beutel, A., Chi, E.H., Dean, J., Polyzotis, N.: The case for learned index structures. In: Proceedings of the 2018 International Conference on Management of Data, pp. 489–504 (2018)
14. Lea, D.: A java fork/join framework. In: Proceedings of the ACM 2000 Conference on Java Grande, pp. 36–43 (2000)
15. Medidi, M., Deo, N.: Parallel dictionaries using AVL trees. J. Parallel Distrib. Comput. **49**(1), 146–155 (1998)
16. Mehlhorn, K., Tsakalidis, A.: Dynamic interpolation search. J. ACM (JACM) **40**(3), 621–634 (1993)
17. Park, H., Park, K.: Parallel algorithms for red-black trees. Theoret. Comput. Sci. **262**(1–2), 415–435 (2001)
18. Paul, W.., Vishkin, U.., Wagener, H..: Parallel dictionaries on 2-3 trees. In: Diaz, Josep (ed.) ICALP 1983. LNCS, vol. 154, pp. 597–609. Springer, Heidelberg (1983). https://doi.org/10.1007/BFb0036940
19. Peterson, W.W.: Addressing for random-access storage. IBM J. Res. Dev. **1**(2), 130–146 (1957)
20. Prokopec, A., Brown, T., Alistarh, D.: Analysis and evaluation of non-blocking interpolation search trees. arXiv preprint arXiv:2001.00413 (2020)
21. Pugh, W.: Skip lists: a probabilistic alternative to balanced trees. Commun. ACM **33**(6), 668–676 (1990)
22. Sleator, D.D., Tarjan, R.E.: Self-adjusting binary search trees. J. ACM (JACM) **32**(3), 652–686 (1985)
23. Sun, Y., Ferizovic, D., Belloch, G.E.: Pam: parallel augmented maps. In: Proceedings of the 23rd ACM SIGPLAN Symposium on Principles and Practice of Parallel Programming, pp. 290–304 (2018)
24. Willard, D.E.: Searching unindexed and nonuniformly generated files in \log\logn time. SIAM J. Comput. **14**(4), 1013–1029 (1985)

Parallel Generation and Analysis of Optimal Chordal Ring Networks Using Python Tools on Kunpeng Processors

Oleg Monakhov$^{(\boxtimes)}$ ⓘ, Emilia Monakhova ⓘ, and Sergey Kireev ⓘ

Institute of Computational Mathematics and Mathematical Geophysics of SB RAS,
Lavrentieva 6, 630090 Novosibirsk, Russia
{monakhov,emilia}@rav.sscc.ru, kireev@ssd.sscc.ru

Abstract. Parallel versions of the reduced exhaustive search algorithm based on the Python tools are implemented to optimize chordal ring networks, which are of practical interest in the design of systems on a chip and supercomputer systems. An analysis of the effectiveness of parallel programs with different numbers of MPI processes on Kunpeng processors was carried out. The speed-up of several parallel computing schemes was experimentally evaluated and analyzed. The large dataset of all optimal chordal networks with numbers of up $6 \cdot 10^4$ nodes was generated for the first time. A preliminary analysis of experimentally obtained dataset has been carried out and the existence of new families of optimal chordal ring networks with analytical descriptions of parameters has been discovered.

Keywords: Optimal chordal ring networks · Parallel algorithm · Discovery of families of graphs

1 Introduction

In [1], Arden and Lee introduced a class of chordal ring networks of degree three as a possible topology for communication networks of multicomputer systems. They investigated a new class in three directions: determining the diameter of graphs, finding the shortest paths for them, and determining the maximum number of vertices for a given diameter. This work aroused great interest in the study of the properties of new structures of communication networks in terms of extending the proposed topology, see, for example, the works [2–11]. Researches were carried out in the following areas: an expanding the range of graph generators, increasing the degree of graph vertices and the degree of its symmetry, considering directed chordal ring networks, including the study of their reliability in case of element failures, building hierarchical structures based on them etc.

Supported by state assignment of ICMMG SB RAS N 0251-2022-0005.

Let us give the basic definitions. An undirected *chordal ring* graph (network), denoted as $C_N(1, -1, s)$, where $3 \leq s \leq N/2$ is an odd number and N is an even number, has a set of vertices $V = Z_N = \{0, 1, ..., N - 1\}$, in which each vertex of the graph i is connected to the vertices $i \pm 1 \pmod{N}$ and each odd vertex i is connected to $i + s \pmod{N}$. The number s is the generator (or chord length) of the graph, N is its order. The graph diameter is the length of the maximum shortest path on the set of all possible pairs of vertices. The diameter estimates the maximum latency, respectively, when executing routing algorithms of a network [12,13]. For a fixed value of N, we call *optimal* a chordal graph $C_N(1, -1, s)$ with the smallest possible diameter for the given N.

Figure 1 shows the optimal chordal network $C_{20}(1, -1, 5)$ of diameter $d = 4$. It should be noted that chordal ring networks of the form $C_N(1, -1, s)$ as bipartite graphs are a subset (with an even number of vertices and ordered removal of edges s or $-s$) of the another widely studied class – circulant graphs of the form $C(N; 1, s)$ [3–5,14].

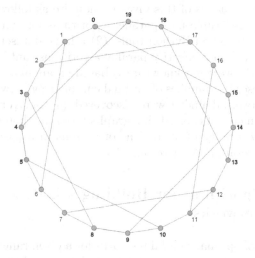

Fig. 1. Chordal ring network $C_{20}(1, -1, 5)$.

In this paper, we study the solution of the optimization problem of finding optimal chordal networks - graphs with a minimum diameter for a given number of vertices (order) of the graph. The authors of [2] obtained the family of extremal graphs with maximal N for a given diameter (in Table 1 it is denoted as f_0). But the problem of minimizing the diameter for a given N is more difficult to solve than maximizing N, because the diameter does not always increase monotonically with increasing value of N. The authors of [2] have also found five other infinite families of graphs with the smallest possible diameter, using dense packing of graphs on the plane. Table 1 shows all these found families of optimal graphs, denoted by us as $f_0 \div f_5$, with descriptions of the parameters of graphs as functions of the diameter d.

Table 1. The families of optimal chordal graphs from [2]

Family [2]	Order of a graph	Generator	$d(mod\ 2)$
f_0	$N = (3d^2 + 1)/2$	$s = 3d$	1
	$N = 3d^2/2 - d$	$s = 3d + 1$	0
f_1	$N = 3d^2/2 - 2d$	$s = 3d - 1$	0
f_2	$N = 3d^2/2 - d - 1/2$	$s = 3d - 2$	1
f_3	$N = 3d^2/2 - 2d - 3/2$	$s = 3d - 4$	1
f_4	$N = 3d^2/2 - 2d + 1/2$	$s = 3d - 4$	1
f_5	$N = 3d^2/2 - 3d - 1/2$	$s = 3d - 6$	1

In this paper, for solution of the optimization problem, the parallel algorithms using three different variants of Python tools for the synthesis of optimal chordal graphs are developed, compared and implemented on cluster of Kunpeng 920 processors. The new features of this investigation are as follows: (1) developed, analyzed and compared three new parallel algorithms for synthesizing optimal chordal ring graphs based on Python tools; (2) a large dataset with $N \leq 60000$ was developed based on new effective parallel algorithms (until now, datasets for chordal ring networks were not known); (3) based on an analysis of the dataset, new analytically described families of optimal chordal ring networks described by functions of the network diameter were discovered; (4) an error was found in the formula for the diameter of chordal ring graphs given in [1]; and (5) the features of using the *igraph* library of operations on graphs in a parallel environment that affect the acceleration are determined.

2   Parallel Algorithms for Building a Dataset of Optimal Chordal Networks

To build a dataset of optimal chordal networks for a given range of nodes, we use and compare three coarse-grain parallel algorithms based on Python tools for parallel processing on task level. Each task consists in synthesis of all optimal chordal networks for a given number of nodes, and the search for all optimal chordal networks for a given range of nodes can be performed in parallel and independently. For synthesis of all optimal chordal networks for a given number of nodes, a reduced sequential brute-force algorithm was developed based on bounds for generators (the bounds are given in Sect. 4). To reduce the time of operations with graphs, a special graph library *igraph* [16] was used. Note that the *igraph* is implemented in the C language and has an order of magnitude higher performance on some operations on graphs than, for example, another popular library for graphs *Networkx* [15].

The first parallel algorithm "Queue" uses the *Queue*() class from Python's *multiprocessing* module [17], which is a first-come-first-served data structure. This class can store simple objects and is useful for exchanging data between processes. The $Queue()$ class of the *multiprocessing* module returns a shared process queue implemented with a pipe and multiple locks/semaphores. Queues are useful when passed as a parameter to a process function to allow it to consume data. Using the $put()$ function, we can push data into the queue, and with $get()$, we can get items from the queues. In the algorithm, the master process spawns a given number of worker processes and two queues *input* and *output*. In the *input* queue, the master process places tasks (the number of vertices for each network for which need to find the optimal parameters) for all worker processes, and in the *output* queue it receives results from worker processes (optimal network parameters). Each worker process takes a task from the head of the *input* queue, independently calculates the optimal parameters, and puts the results of the computation at the end of the *output* queue. The algorithm ends when the master process receives the results for all tasks.

The second parallel algorithm "Pool" uses the $Pool()$ class of the *multiprocessing* module [17] and a master process creates an object that manages a pool with given number of worker processes to which tasks are submitted. The worker process pool supports asynchronous execution of tasks and has a parallel implementation. In the algorithm a multiprocessing pool in Python is used to execute a given function by applying that function to each element in parallel using the Pool $map()$ method. For the synthesis of optimal chordal ring networks, the specified range of the number of vertices is evenly distributed among parallel processes (data parallelism). Each worker process independently executes its tasks, and after parallel processing all tasks the master process receives the results of execution.

The third parallel algorithm "MPI" is based on MPI for Python package (mpi4py) [18], which provides Python binding to the Message Passing Interface (MPI) standard, allowing Python applications to use multiple processors in clusters. The package builds on the MPI specification and provides an object oriented interface resembling the MPI-2 C++ bindings. The third parallel algorithm creates *size* processes, and each process with rank r independently executes tasks with node numbers $\{nmin+r, nmin+r+size, nmin+r+2*size, ...\}$, based on a simple cyclic (round-robin) procedure for a given range of nodes $(nmin, nmax)$ of chordal networks. This rather simple algorithm does not require any interaction and synchronization between processes, except for synchronization of the end of computations.

3 Experimental Results with Parallel Algorithms

Computational experiments were carried out in order to compare the results of various implementations of parallel algorithms for the synthesis of optimal chordal ring networks. The results of the synthesis were compared in terms of the speedup and efficiency of the synthesis of all optimal networks for a given range of the number of vertices and for a given number of processes. Computational experiments were mainly carried out on an one node of cluster of ICMMG SB RAS with the following characteristics: 2 × 64-core Kunpeng 920

2.6 GHz processors (128 cores per node), openEuler v20.09 operating system, gcc v.9.3.1 compiler, OpenMPI 4.1.0 library, Python 3.8.5 with libraries mpi4py 3.1.4, igraph 0.10.4, networkx 3.0. The implementations of the considered synthesis algorithms were used to construct a dataset of graphs with up to $6 \cdot 10^4$ vertices. An experimental comparison of the algorithms was carried out when obtaining the results of the synthesis of networks with the number of vertices from 5000 to 10120.

During the experiments, we found that the standard import of the igraph module leads to the fact that this module is loaded on only one processor core and all other processes send requests to this core. To overcome this bottleneck, we have used the Python's standard utility OS module with $os.sched\_ setaffinity()$ method to set a per-process affinity mask that specifies the core or set of cores the process can run on.

Figure 2 and Fig. 3 demonstrate the speedup and efficiency obtained on one node for a given number of processes with respect to a sequential program (one process) for different variants of implementation for the parallel synthesis algorithm. The following set of configurations were considered for MPI, Pool and Queue programs with Pr in $\{1, 2, 4, 8, 16, 32, 64, 128\ \}$ (Pr is the number of processes). Figure 2 and Fig. 3 show that the maximum speedup and maximum efficiency on one node is obtained with the MPI program. For example, we get speedup equal to 110.1 with efficiency 0.86 for 128 processes. The greater efficiency and acceleration of the MPI program is explained by the absence of overhead costs for organizing of a job queue or a pool of processes. An another advantage of the MPI program is that it can be executed on several cluster nodes, unlike the other two programs. For example, with the MPI program we get speedup equal to 222.1 with efficiency 0.86 for two nodes with 256 processes and speedup equal to 319 with efficiency 0.83 for three nodes with 384 processes.

4   Analysis of the Dataset of Optimal Chordal Networks

For development of the synthesis algorithm we used the following options to reduce the search for optimal graphs. First, when determining the diameter of a chordal graph, due to its symmetry, one can use finding the length of the maximum path in a graph (diameter) from one a vertex, for example, zero. This reduces the diameter calculation time proportionally to N. Second, an another option to reduce the search time and, accordingly, increase the size of the base of optimal graphs obtained is to use parallel search algorithms. This option and different versions of its realization have been considered in Sects. 2 and 3 in detail.

Using the search algorithm with the options indicated, we have built a dataset of optimal chordal graphs. For a given N, all generators $3 \le s \le N/2$ were enumerated and the graph(s) with the minimum diameter was (were) determined. Table 2 shows a fragment of the resulting dataset of optimal chordal ring networks (N, s, d) for $130 \le N \le 166$. The corresponding cells contain values of order N, generators s and the minimum diameter d of the graph. The full version of the dataset for $N \le 60000$ will be available in github.

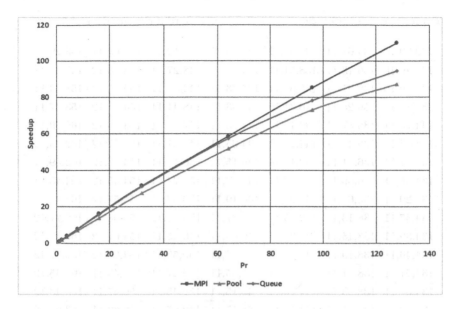

Fig. 2. The speedup obtained for a given number of processes for different variants of the parallel synthesis algorithm

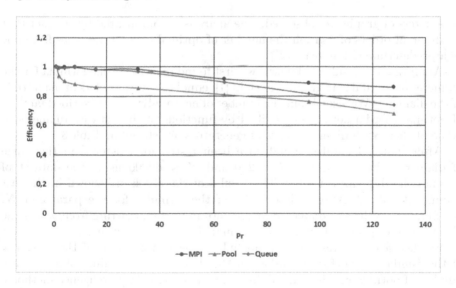

Fig. 3. The efficiency obtained for a given number of processes for different variants of the parallel synthesis algorithm

Figure 4 shows a plot of values of diameter d versus N ($N \leq 24000$) and s of optimal chordal networks for a part of dataset. Interesting dependencies between

Table 2. Fragment of the dataset of optimal chordal ring networks (N, s, d).

130,39,10	134,61,11	138,41,11	144,19,11	148,19,11	152,69,11	158,29,11
130,49,10	136,13,11	138,51,11	144,39,11	148,27,11	154,13,12	158,35,11
132,29,10	136,21,11	138,61,11	144,43,11	148,33,11	154,15,12	158,43,11
134,13,11	136,25,11	140,31,10	144,53,11	148,41,11	154,21,12	158,47,11
134,15,11	136,37,11	142,13,11	144,63,11	148,65,11	154,23,12	160,29,11
134,17,11	136,41,11	142,15,11	144,65,11	148,67,11	154,27,12	162,29,11
134,21,11	136,51,11	142,17,11	146,15,11	150,17,11	154,43,12	162,59,11
134,25,11	136,59,11	142,19,11	146,17,11	150,27,11	154,45,12	164,45,11
134,29,11	136,61,11	142,21,11	146,19,11	150,33,11	154,57,12	164,49,11
134,37,11	138,13,11	142,25,11	146,27,11	150,41,11	154,65,12	166,17,12
134,39,11	138,15,11	142,31,11	146,43,11	150,45,11	154,67,12	166,19,12
134,49,11	138,19,11	142,39,11	146,53,11	150,55,11	154,69,12	166,37,12
134,51,11	138,21,11	142,63,11	146,55,11	152,27,11	156,43,11	166,45,12
134,57,11	138,25,11	144,15,11	148,15,11	152,45,11	156,57,11	166,49,12
134,59,11	138,31,11	144,17,11	148,17,11	152,67,11	156,69,11	166,61,12

parameters of graphs are observed. The analysis is complicated by the fact that for different orders of a graph the number of optimal generators varies from one to a predetermined number

Analyzing the resulting dataset, we first found an error in the formula for the diameter of chordal ring graphs, given in [1], equal to several orders of magnitude of the diameter. We also obtained a number of new analytically described families of optimal chordal networks described by functions of the network diameter d. Some of them with different types of generators are shown in Table 3.

After that, all families found have been tested for existence in the range of diameters from $d = 4$ to $d = 200$ and $N \leq 60000$ using the dataset of optimal chordal ring networks. The results of the check are shown in the last column of Table 3. After a description of the formulas for the parameter N, the minimum values of diameters (d_{min}) are indicated, starting from which the obtained descriptions of families of optimal chordal networks exist.

Further scientific research will include a theoretical proof of the existence of the families found for any values of the diameter and finding other infinite families of optimal graphs using data mining and artificial intelligence methods.

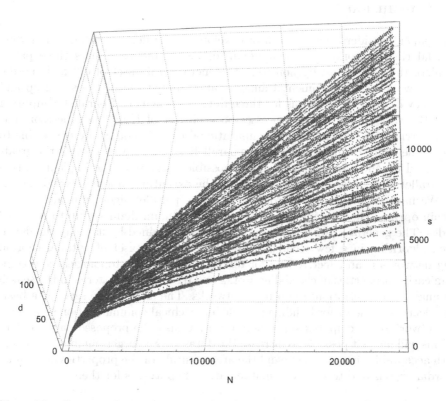

Fig. 4. The diameter d versus number of nodes N and generators s of optimal chordal ring networks

Table 3. New families of optimal chordal networks

Family	Order of a graph	Generator	$d(mod\ 2)$	d_{min}
k_1	$N = 3d^2/2 - 4d,$	$s = 3d + 1$	0	10
k_2	$N = 3d^2/2 - 4d - 2,$	$s = 3d + 1$	0	12
k_3	$N = 3d^2/2 - 7d,$	$s = 3d + 1$	0	30
k_4	$N = 3d^2/2 - 7d - 4,$	$s = 3d + 1$	0	34
k_5	$N = 3d^2/2 - 3d + 3/2,$	$s = 3d$	1	7
k_6	$N = 3d^2/2 - 3d - 1/2,$	$s = 3d$	1	7
k_7	$N = 3d^2/2 - 6d + 5/2,$	$s = 3d$	1	15
k_8	$N = 3d^2/2 - 6d - 3/2,$	$s = 3d$	1	19
k_9	$N = 3d^2/2 - 5d,$	$s = 3d - 1$	0	14
k_{10}	$N = 3d^2/2 - 8d,$	$s = 3d - 1$	0	38

5 Conclusion

The paper considers the problem of developing parallel algorithms for finding chordal ring networks with a minimum diameter and represents three parallel algorithms based on Python tools to generate a dataset of optimal chordal ring networks. An experimental analysis and comparison of the speed-up and efficiency of parallel programs for three different schemes of parallel computing with the various number of processes on a cluster of Kunpeng processors has been carried out. The results of computational experiments made it possible to determine which configurations for the parallelization schemes provide the greatest speed-up and efficiency among the possible variants of the implementation of parallel algorithms and allow to reduce the execution time by several times.

We have shown that the proposed effective parallel algorithms allow us to obtain optimal chordal ring networks with a minimum diameter up to $N \leq 10^5$ nodes. The generators of optimal networks were obtained by us using the developed algorithm in a few seconds. The constructed dataset of optimal chordal ring networks is an effective tool for further studying the topological and communicative properties of chordal networks and discovering patterns in the search for analytical descriptions of optimal networks. The dataset can serve as a basis for selecting elements for building reliable hierarchical communication structures for networks on a chip and supercomputer systems. The proposed parallel algorithms will make it possible to increase the dataset for a larger number of vertices with acceptable time costs, expand the area of study of the properties of optimal chordal graphs and test various mathematical hypotheses for them.

References

1. Arden, B.W., Lee, H.: Analysis of chordal ring network. IEEE Trans. Comput. **C-30**(4), 291–295 (1981)
2. Morillo, P., Comellas, F., Fiol, M.A.: The optimization of Chordal Ring Networks. Commun. Technol., Eds. Q. Yasheng and W. Xiuying. World Scientific, 295–299 (1987)
3. Bermond, J.-C., Comellas, F., Hsu, D.F.: Distributed loop computer networks: a survey. J. Parallel Distrib. Comput. **24**, 2–10 (1995)
4. Hwang, F.K.: A survey on multi-loop networks. Theoret. Comput. Sci. **299**, 107–121 (2003)
5. Monakhova, E.A.: A survey on undirected circulant graphs. Discr. Math. Algorithms Appl. **4**(1), 1250002 (2012)
6. Pedersen, J.M., Riaz, T.M., Madsen, O.B.: Distances in generalized double rings and degree three chordal rings. In: IASTED International Conference on Parallel and Distributed Computing and Networks (IASTED PDCN2005), pp. 153–158. Austria (2005)
7. Parhami, B.: Periodically regular chordal rings are preferable to double-ring networks. J. Interconnection Netw. **9**(1), 99–126 (2008)
8. Farah, R.N., Chien, S.L.E., Othman, M.: Optimum free-table routing in the optimised degree six 3-modified chordal ring network. J. Commun. **12**(12), 677–682 (2017)

9. Chen, S.K., Hwang, F.K., Liu, Y.C.: Some combinatorial properties of mixed chordal rings. J. Interconnection Netw. **4**(1), 3–16 (2003)
10. Gutierrez, J., Riaz, T., Pedersen, J., Labeaga, S., Madsen, O.: Degree 3 networks topological routing. Image Process. Commun. **14**(4), 35–48 (2009)
11. Ahmad, M., Zahid, Z., Zavaid, M., Bonyah, E.: Studies of chordal ring networks via double metric dimensions. Math. Probl. Eng., Article ID 8303242 (2022). https://doi.org/10.1155/2022.d303242
12. Monakhova, E.A., Monakhov, O.G., Romanov, A.Y.: Routing algorithms in optimal degree four circulant networks based on relative addressing: comparative analysis for networks-on-chip. IEEE Trans. Netw. Sci. Eng. **10**(1), 413–425 (2023)
13. Monakhov, O., Monakhova, E., Romanov, A., Sukhov, A., Lezhnev, E.: Adaptive shortest path search algorithm in optimal two-dimensional circulant networks: implementation for networks-on-chip. IEEE Access **8**, 215010–215019 (2021)
14. Huang, X., Ramos, A.F., Deng, Y.: Optimal circulant graphs as low-latency network topologies. J. Supercomput. **78**, 13491–13510 (2022). https://doi.org/10.1007/s11227-022-04396-5
15. Platt, E.: Network Science with Python and NetworkX Quick: Start Guide. Packt, Birmingham, UK (2019)
16. igraph - The network analysis package. https://igraph.org/. Accessed 14 May 2023
17. Zaccone, G.: Python Parallel Programming Cookbook. Packt, Birmingham, UK (2015)
18. Dalcin, L., Paz, R., Storti, M., D'Elia, J.: MPI for Python: performance improvements and MPI-2 extensions. J. Parallel Distrib. Comput. **68**(5), 655–662 (2008). https://doi.org/10.1016/j.jpdc.2007.09.005

Combinatorial Aspect of Code Restructuring for Virtual Memory Computer Systems Under WS Swapping Strategy

Stepan Vyazigin[✉] [iD] and Madina Mansurova[iD]

Al-Farabi Kazakh National University, Almaty, Kazakhstan
wismas1996@gmail.com

Abstract. This paper presents innovative findings on the restructuring of code for virtual memory systems operating under a working set swapping strategy. Despite extensive research spanning five decades and numerous studies dedicated to restructuring, the persisting absence of definitive solutions has motivated this inquiry. The NP-hard problem of code block relocation across virtual memory pages to minimize cost function reflects a core challenge inherent to the problem. For ill-defined programs, many practical cluster-based solutions lack a quantifiable approximation error to the unknown optimal or ε-optimal solution. This paper elucidates the computational process by offering a geometric interpretation, enabling the construction of a combinatorial mathematical model of the restructuring process. This model incorporates both functional elements and constraints to define acceptable solutions. The unique aspects of the model provide a foundation for subsequent research aimed at designing an algorithm that delivers an optimal or ε-optimal solution to the original problem, with some algorithmic details discussed herein. The model also paves the way for the development of a swift, cost-effective working set-like swapping algorithm, amplifying the applicability of the results obtained.

Keywords: working set · swapping strategy · shared virtual memory system · combinatorial space · restructuring · ill-defined program code

1 Introduction

Program transformations, such as program restructuring [1–9] and refactoring [10, 11], or other types of code reorganization are widely recognized for their positive impact on program behavior, including code locality. This impact is especially noticeable for ill-structured, frequently run programs, prompting researchers to strive for optimal, or at the very least, practically acceptable solutions when implementing code transformations. Importantly, these transformations must not compromise program properties such as correctness and locality.

In this context, it's important to note that many studies on restructuring, including our research focus, primarily employ clustering techniques [2, 6], which, despite yielding improved experimental outcomes, provide only approximate solutions of unknown

V. Malyshkin (Ed.): PaCT 2023, LNCS 14098, pp. 136–147, 2023.
https://doi.org/10.1007/978-3-031-41673-6_11

accuracy. Specifically, this paper emphasizes the problem of program reorganization for virtual memory systems, where the clustering technique offers the potential to enhance program behavior, albeit with an unknown approximation to the exact solution. We are particularly interested in the combinatorial aspect of the research, and our approach does not rely on a universally accepted method such as clustering.

Virtual memory, a common memory organization concept, extends beyond the typical memory hierarchy of desktop PCs and multicomputer systems like shared virtual memory (SVM) systems. It also serves as an interface between the central and graphics processors, as exemplified in Compute Unified Device Architecture (CUDA) [12].

Our results, distinct from those aforementioned, pertain to the combinatorial aspect of restructuring. We believe they will prove useful in studying similar issues in multicore and multiprocessor systems, such as SVM-like multicore systems. Concerning the working set (WS) swapping strategy, it's worth mentioning that although WS strategies are often used as a theoretical basis for research or for comparative or auxiliary purposes, they are generally considered costly to implement. However, for programs with a block structure, our findings suggest the potential to construct a rapid, cost-effective WS-like swapping algorithm.

One common approach to improving program structure is program refactoring [10, 11]. Despite the superficial similarity, code restructuring and code refactoring bear distinct differences. Code refactoring techniques mainly aim at transforming object-oriented code without considering how the transformation aligns with the optimal code version. In contrast, our work focuses on instruction-level code where refactoring methods are not applicable. We face two intertwined issues: diminished system performance due to excessive page faults and deteriorated code locality. The initial question is identifying the specific code items causing these issues. Through initial code experimentation using the WS swapping algorithm, we can identify a subset of items suitable for restructuring, which we designate as situation A. Alternatively, situation B involves identifying a predefined set of interacting code items for restructuring.

2 Conceptual Level. The Initial Statement of the Main Problem of Program (Code) Optimization

Let us consider a code (or program) consisting of n interacting blocks, $b_1, b_2, \ldots, b_n$ (abbreviated as 1, 2, ..., n), that have been pre-identified and distributed across p pages $S_{g_1}, S_{g_2}, \ldots, S_{g_p}$ of virtual memory. For simplicity, we denote these pages as $S_1, S_2, \ldots, S_p$. The code execution may cause a problem due to an unexpectedly high number of page faults, significantly impacting the system because of the code's ill-defined structure, which in turn reduces the performance of both the code (program) and the system itself. Such adverse behavior can also result from code segments being written by different authors from disparate programming teams at different times. Under the previously mentioned situation B, any reference from the residential set during code execution should be to one block from $\{b_1, b_2, \ldots, b_n\}$ only. For situation A, this restriction is not mandatory, and we will elaborate on this later.

Let v_r denote the length of the r-th page $r = 1, 2, \ldots, p$ and let l_i be the length of block $i = 1, 2, \ldots, n$. This notation indicates that the system supports multidimensional page sizes. Here, blocks refer to code components such as subroutines, linear segments of code, independent interacting programs, application data blocks, etc. The distribution of blocks $b_1, b_2, \ldots, b_n$ across pages $S_1, S_2, \ldots, S_p$ is predetermined, represented via a Boolean matrix $x = (x_{ri})_{p \times n}$, where an element $x_{ri} = 1$ if block i belongs to page r and $x_{ri} = 0$ otherwise. We denote all such matrices by X.

Consider that random data, denoted by $D = \{\theta\}$, and the swapping strategy also impact the value of the functional. Assuming that after code execution with any given $\theta \in D$, the reference string of blocks (pages) corresponding to θ is available (Fig. 1, 2). This paper chooses the working set (WS) strategy as its swapping strategy. The residential set R of pages in the main memory at any moment t of a program run coincides with the working set of pages at the moment t, denoted by the universally accepted notation $W(k, t)$ [3]. We regard the block analogy of the working set as a control state (c.s.) of the code (program), using q as the corresponding notation for c.s.

It is important to note that there are natural constraints (a)–(c) on the matrix $x \in X$, which we will conceptually describe below before presenting formal correlations and details:

Functionals: For the main problem, we consider the mathematical expectation (expected value) of the number of page faults per program (code) run as the functional. For the auxiliary problem, the functional is the mean value of page faults for h $\geq$ 1 program (code) runs.

Constraint (a): This constraint implies that the total length of the blocks allocated to any page does not exceed the length of that page.

Constraint (b): This constraint states that any block of the program (code) belongs to only one page of the program (code).

Constraint (c): This constraint requires that the total length of any working set generated during the program (code) execution does not exceed a system constant known in advance.

Remember, these constraints (a)–(c) are defined by the matrix $x = (x_{ri})_{p \times n}$, which determines the distribution of blocks $b_1, b_2, \ldots, b_n$ across pages $S_1, S_2, \ldots, S_p$.

Despite the abundance of papers dedicated to program restructuring, particularly for our interpretation of the problem [2–6], no definitive solution or appropriate model for reorganization to achieve exact (optimal) solutions with the aforementioned functional exists.

3   Control State of the Program. Working Set Generated by Control State q and Matrix $x = (x_{ri})_{p \times n}$. The Set of Control States Q

When considering the concept of the working set, we typically refer to two related notations derived from P. Denning: $W(t - \tau, t)$ with a window size τ, and $W(k, t)$ with parameter k. In the latter case, an integer $k \geq 1$ can also be interpreted as the window size. In the first scenario, the working set refers to the pages of a program referenced within the interval $[t - \tau, t)$ of virtual time. In our case, we have adopted the variant $W(k, t)$ as the working set. Considering the control state (c.s.) q_t of the program at moment t, let us define it as the set of program blocks referenced in the last k moments prior to time t. Therefore, the control state (c.s.) of the program (code) at moment t essentially represents the block analogy of the working set for pages at that same moment.

It's important to note the role of the Boolean matrix $x = (x_{ri})_{p \times n}$, which, as previously mentioned, defines the program's structure, i.e., the distribution of blocks $b_1, b_2, \ldots, b_n$ across pages $S_1, S_2, \ldots, S_p$. As we established, this matrix x must comply with constraints (a)–(c), and all matrices of this type constitute the set X. Ultimately, among the matrices in X, an optimal matrix should be identified that assigns an optimal structure to the code (program) based on the aforementioned functional.

3.1   Reference Strings to Pages and Blocks. Control State q_t. Working Set $R(q_t, x)$

Examining the reference string for one program run (Fig. 2) and the corresponding frame divided into k cells beneath it, which incrementally moves from left to right along the time axis t, we should concurrently direct our attention to Fig. 1.

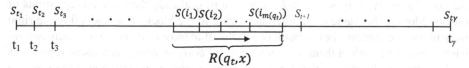

Fig. 1. The reference string to the program's pages for a single program run, where the working set R(q_t, x) is derived, corresponds to W(4, t) with k = 4, under t = t1, t2, ..., tγ.

In Fig. 1 and 2, moments t1, t2, t3, ..., tγ represent instances of references, starting from t0 and ending at tγ, which is the final moment for the run with the random data θ ∈ D. In Fig. 1, over the time axis t, the notations St, t = 1, 2, 3,..., γ represent the pages (or their numbers) that have been referenced during the execution of the program for a specific θ ∈ D, given k = 4, and a fixed x ∈ X. For Fig. 2, notations $i(t_j), j = 1, 2, \ldots, t\gamma$, are the numbers of blocks corresponding to the page numbers in Fig. 1 that have been referenced during the execution of the program for the same θ ∈ D, given k = 4, and the same x ∈ X.

The contents of the frame in Fig. 2 are blocks, potentially with repetitions, which form the control state (c.s.) q_t at moment t. The contents of the frame in Fig. 1 are page numbers, also potentially with repetitions, which form the working set $R(q_t, x)$ generated by the c.s. q_t and matrix $x = (x_{ri})_{p \times n}$ at moment t. Each page $S(i_j)$ in the frame contains block i_j of c.s. q_t, where $j = 1, 2, \ldots, m(q_t)$.

It's For $R(q_t, x)$, an important condition holds: every page within $R(q_t, x)$ contains at least one block of the c.s q_t. In our case, no other type of working set exists. Thus, as a multiset, $R(q_t, x)$ is $\{S(i_1), S(i_2), \ldots, S(i_{m(q_t)})\}$, but the actual working set corresponding to $R(q_t, x)$ should not contain repeated pages. Let $R(q, x)$ denote the working set without page repetitions corresponding to $R(q_t, x)$. Note, it's easy to perform this transformation by eliminating the repetitions in q_t to transition from $R(q_t, x)$ to $R(q, x)$.

Proceeding from Fig. 2 it is easy to notice that the next c.s. q_{t+1} forms as $q_t \cup \{i\}$, i.e. $q_{t+1} = \{i_2, i_3, \ldots, i_{m(q_t)}, i\}$. Such kind of event, i.e. reference from c.s. q_t to the block i we may denote as $q_t \rightarrow i$ and number i becomes available by the end of processing c.s. q_t.

Thus, as we can see, a content of the frame at any moment t with repetitions of blocks numbers coincides with a control state (c.s.) of the program at moment t, the denotation

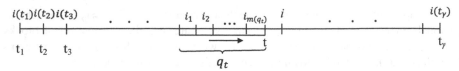

Fig. 2. Reference string to blocks of a program for one run of the program, where for any t of the process is derived c.s. q_t

for it is $q_t = \{i_1, i_2, \ldots, i_{m(q_t)}\}$ (see above and Fig. 2), in contrast with corresponding denotation for c.s. without repetitions, i.e. $q = (i_1, i_2, \ldots, i_{m(q)})$. Besides we can omit index t from q_t since for us there are no differences between $\{i_1, i_2, i_3\}$ and $\{i_2, i_1, i_3\}$ or $\{i_3, i_1, i_2\}$ and so on and we will write instead of all of them, the ordered record (i_1, i_2, i_3), where $i_1 < i_2 < i_3$. And the same for c.s. $q = (i_1, i_2, \ldots, i_{m(q)})$ takes place the correlation: $i_1 < i_2 < \ldots < i_{m(q)}$, where in q already there are no repetitions of blocks. Moving the frame along reference string to the blocks from the left to the right we will receive some of the c.s. but it is possible that most of them are multisets and we have considering both of them, i.e. a multiset and corresponding set without repetitions as the same set. In spite of that for q_t, in contrast with q, we introduced the separate refinement of the denotation, namely $q_t = \{i_1, i_2, \ldots, i_{m(q_t)}\}$ and treated q_t as a multiset. Having excluded from found q_t the repetitions of block's numbers and then if q_t is new, that checking, by way of comparing q_t with q from the set $Q = \{q\}$, that has formed for the current moment t, then q_t has to be included into Q, as new q. Thus, by way of new runs of the program (code) by and by we reach a situation when the set Q does not change. And finally the set $Q = \{q\}$ consist of different q, where in any q there are no repetitions of blocks and Q does not change even after $\lambda \geq 1$ additional runs.

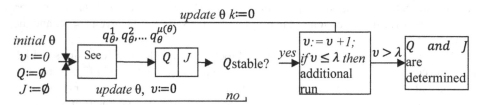

Fig. 3. Sets Q and J generation

Here at the Fig. 3, the set $\left\{q_\theta^1, q_\theta^2, \ldots, q_\theta^{\mu(\theta)}\right\}$ is a multiset of control states, are obtained under one run of the code, corresponding a separate $\theta \in D$, and anyone new of them without repetitions will be included into Q. Each subsequent run brings us closer to a stable Q. The condition "Is Q stable?" in Fig. 3, when answered affirmatively for the first time, indicates that Q was unstable in previous iterations but has become stable in the current iteration, showing no difference from Q in the preceding iteration. If υ is less than or equal to λ, the process must follow the upper line, update θ, and proceed according to the scheme.

The set J is the collection of block numbers i that occur during the transition $q_t \to i$ as the frame moves along the time axis in Fig. 2. After at least λ additional runs, by reassigning elements of J new numbers from $1, 2, \ldots$ to n, we will obtain the set of blocks that need restructuring, which is suitable for further examination.

However, a question arises: what if, even after λ or more additional runs, a new control state is discovered? This would suggest that the choice for λ was overly optimistic and that λ needs to be reassessed, necessitating a restart of the process.

The situation described above corresponds to situations A and B (see the end of point 1 in the Introduction), but for situation B, the set J is ignored. Note that in situation B, the set J is predefined.

3.2   Correlations Between c.s. q_t and c.s. q_{t+1}

Here, the c.s. q_{t+1} (see Fig. 2) forms from subset of q_t, including the case q_t itself, which has to be join with $\{i\}$ and then the next correlations hold: $q_{t+1} \subset q_t$ or $q_t \subset q_{t+1}$, or $q_{t+1} = q_t$. Indeed, c.s. $q_t = \{i_1, i_2, \ldots, i_{m(q_t)}\}$ and $q_{t+1} = \{i_2, i_3, \ldots, i_{m(q_t)}, i\}$ (see middle fragment of Fig. 2), then we have several steps to continue:

I. If $i_1 \in q_{t+1} = \{i_2, i_3, \ldots, i_{m(q_t)}, i\}$ and $i \notin \{i_2, i_3, \ldots, i_{m(q_t)}\}$, then $q_t \subset q_{t+1}$, (see Fig. 4.a).
   The next situation is

II. if $i_1 \in q_{t+1}$ and $i \in \{i_2, i_3, \ldots, i_{m(q_t)}\}$ then $q_t = q_{t+1}$ (see Fig. 4.b).
   And further,

III. if $i_1 \notin \{i_2, i_3, \ldots, i_{m(q_t)}\}$ and $i \notin \{i_1, i_2, \ldots, i_{m(q_t)}\}$ then $q_t \not\subseteq q_{t+1}$(see Fig. 4.c).
   Here $q_t = \{i_1, i_2, \ldots, i_{m(q_t)}\}$ and $q_{t+1} = \{i_2, i_3, \ldots, i_{m(q_t)}, i\}$

IV. If $i_1 \notin \{i_2, i_3, \ldots, i_{m(q_t)}\}$ and $i \in \{i_2, i_3, \ldots, i_{m(q_t)}\}$ then $q_{t+1} \subset q_t$ (see Fig. 4.d), where $q_t = \{i_1, i_2, \ldots, i_{m(q_t)}\}$, $q_{t+1} = \{i_2, i_3, \ldots, i_{m(q_t)}\}$.

Steps I, II, IV are standard, but for step III we may consider, that while intermediate node q go by the process, the time moment equal $t + 1/2$ (see Fig. 4.c).

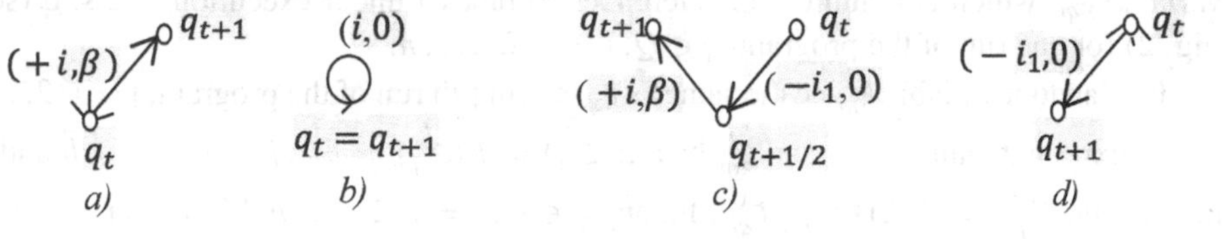

Fig. 4. Variants of movement in a bipolar combinatorial space

At the fragments of Fig. 4 we can notice the arc (q_t, q_{t+1}), and the arc $(q_t, q_{t+1/2})$ and the arc $(q_{t+1/2}, q_t)$, each having the weight $(\pm i, \beta)$, where $\beta \in \{0, 1\}$. Here $\pm i$ means a movement up or down that either enriches c.s. q_t (see Fig. 4.a), with i moving up or impoverishes the c.s. q_t (see Fig. 4.d), by way of i_1 (movement down) or sequentially both of them (Fig. 4.c). A number i without sign (Fig. 4.b) means the situation $q_t = q_{t+1}$. Parameter β equal 1 or 0 and related the event, when page fault happens or not under corresponding movement from c.s. q_t to c.s. q_{t+1}.

3.3 Final Notices to Determine the Set of Control States Q

After conducting multiple runs of the program under different $\theta \in D$ and repeatedly identifying c.s. q_t, once we reach the situation where the set Q does not change, and after the next additional $\lambda \geq 1$ runs the set Q remains the same, we will consider the set Q as defined (see Fig. 3, also refer to the conclusion of point 4.2, step 1.). For a moment, let's assign a special c.s. $q_0 = \varnothing$ to the set Q, which initially corresponds to the starting point of the process. The c.s. q_0 may also occur at later moments when our program unexpectedly offloads from main memory and is reactivated after a while as if it's running from the start (cold start). Another start is a warm start, where the system ensures the computational process, including the c.s., is restored to the state just before the program was offloaded, allowing the program to bypass the wait time in secondary memory and continue the process as if no offload occurred. Going forward, we propose treating any such event as a warm start (restart), considering it as an additional page fault, which will be accounted for in additional terms in expressions (1), (2) for the functional of the main and auxiliary problem (see details in point no. 4).

4 Random Variables. Describing Functionals and Constrains of the Main and Auxiliary Problems

In this point, we start from finding expressions both the functional of main and auxiliary problem and expressions for corresponding constrains for matrix $x = (x_{ri})_{p \times n}$. As well, it is useful to discuss about geometric aspect of our approach to the initial problem. Further we are considering, that sets $Q, \widehat{Q}$ are determined.

4.1 Random Variable ξ_{qi} and the Function $\delta_{qi}(x)$, Where $q \in Q, i = 1, 2, \ldots, n$

Drawing upon the information provided earlier, let's revisit the topic of the random variable ξ_{qi}, which is a number of references to block i under execution of c.s. q (see Fig. 2) for one run of the program, $q \in Q, i = 1, 2, \ldots, n$.

Let random variable ξ_{qi}^{j} be the same as ξ_{qi} but in j-th run of the program, j = 1, 2, ..., h. Let expected value of ξ_{qi} and ξ_{qi}^{j} be a $E(\xi_{qi}) = E(\xi_{qi}^{j}) = E_{qi}, j = 1, 2, \ldots, h$ and a mean value $E_{qi}^{(h)} = (1/h) \cdot \sum_{j=1}^{h} \xi_{qi}^{(j)}$, for any $q \in Q, i = 1, 2, \ldots, n$. More over, we have suggested that a series of h $\geq$ 1 runs are carried out with our program (code). After each run, denoted by number j, the variables ξ_{qi}^{j} $(q \in Q, i = 1, 2, \ldots, n)$ have been calculated. Remember, eventually concerning ξ_{qi}^{j} we have to calculate mean value $E_{qi}^{(h)}$ for which, convenient for us, a recurrent formula is proposed beneath (see point no. 4.2), for all $q \in Q, i \in \{1, 2, \ldots, n\}; i \notin q$. Thus the values $E_{qi}^{(h)}$, $(q \in Q, i \in \{1, 2, \ldots, n\}; i \notin q)$ is having been calculated after h runs of the program.

An intriguing question to address is whether the reference to block i under the execution of c.s. q triggers a page fault or not (as depicted in the middle section of Fig. 2). Such an event, as illustrated in the middle section of Fig. 2, can occur anywhere along the axis t during the execution of the program, even within a single run. Let's denote this event, by analogy with $q_t \rightarrow i$, as $q \rightarrow i$, and under a fixed x, the event

$q \rightarrow i$ produces the same response regardless of where along the axis t it occurs. The response to this question is given by a function that depends on the c.s. q and the matrix $x = (x_{ri})_{p \times n}$. Namely the function $\delta_{qi}(x)$:

$$\delta_{qi}(x) = \begin{cases} 0, \ \textit{if block } b_i \in S \in R(q, x) \\ 1, \ \textit{otherwise} \end{cases}$$

where S is denotation of a page from $R(q, x)$. The value $\delta_{qi}(x) = 0$ corresponding the absence of the page fault under event $q \rightarrow i$, i.e. the value $\delta_{qi}(x) = 0$, if block i belongs some page S from $R(q, x)$. Otherwise, the value $\delta_{qi}(x) = 1$ corresponding to the page fault. Of course, if block $i \in q$ then it has to be $\delta_{qi}(x) \equiv 0$ for any $x \in X$. Moreover if matrix $x = (x_{ri})_{p \times n}$ is fixed then it does not matter at what moment and under which runs of the program happen the event $q \rightarrow i$. In any case, between the two possible outcomes - a page fault either occurs or does not occur for all events $q \rightarrow i$, meaning the answer is consistent for all events $q \rightarrow i$ given q, i, and a fixed $x \in X$. In other words, the same outcome applies to all events $q \rightarrow i$, regardless of where along the axis t the event takes place (refer to Fig. 2, point no. 3). It's important to note that any matrix $x \in X$ adheres to restrictions a)–c) mentioned earlier (also refer to point no. 4.2 correlations (3)–(6)).

Calculation of the $\delta_{qi}(x)$ can be done the way

$$\delta_{qi}(x) = \begin{cases} 0, \ if \ \sum_{j=1}^{m(q)} \sum_{r=1}^{p} x_{ri} \cdot x_{rij} \geq 1, q \neq q_0 \\ 1, \ \text{otherwise} \end{cases}$$

and it is easy to notice that if block $i \in q$ then $\delta_{qi}(x) = 0$ for any $x \in X$.

4.2 Functionals $F^0(x)$ and $F^{(h)}(x)$ and Constraints for $x \in X$

From now on let us remove c.s. q_0 from Q, and let $\theta \in D$. Taking into account content of the points no. 3 and point no. 4.1 a total number of page faults for one run of the program, let it be random function $\xi_\theta(x)$

$$\xi_\theta(x) = \sum_{q \in Q.} \sum_{i=1}^{n} \xi_{qi} \cdot \delta_{qi}(x) + \sum_{i=1}^{n} \xi_{q_0 i},$$

then for the functional of the main problem, which has to be minimize we have

$$F^0(x) = \sum_{q \in Q.} \sum_{i=1}^{n} E_{qi} \cdot \delta_{qi}(x) + \sum_{i=1}^{n} E_{q_0 i} \rightarrow \min_{x \in X} \qquad (1)$$

It worth to note that in expression for ξ above, the any value ξ_{qi} does not depends on matrix $x \in X$ and in contrast, the function $\delta_{qi}(x)$ depends on given $q \in Q$ and i and $x \in X$, and not depends on, where at axis t the random event $q \rightarrow i$ happens.

For the functional $F^{(h)}(x)$ of auxiliary problem holds

$$F^{(h)}(x) = \sum_{q \in Q.} \sum_{i=1}^{n} E_{qi}^{(h)} \cdot \delta_{qi}(x) + \sum_{i=1}^{n} E_{q_0 i}^{(h)} \rightarrow \min_{x \in X} \qquad (2)$$

It is interesting to note that value $E_{qi}^{(h)}$ from (2) can be assigned as a weight to the edge (arc), which connects the node q and the node $q \cup i$ of the Boolean, where the function $\delta_{qi}(x) = 1$ (movement up). Otherwise, i.e., if the function $\delta_{qi}(x) = 0$ this edge (arc) has to be weighted as zero. It may helps to calculate the value of the functional $F^{(h)}(x)$ for fixed $x \in X$, but, actually, not necessary do such kind of modification the edges of the Boolean. It will be enough, if weight of the any edge (arc) under consideration, directed upward, equals 1 or 0, depending on page fault happen or not (remember the function $\delta_{qi}(x)$) under transition from adjacent nodes, as already mentioned above. Then, in corresponding cases, the values $E_{qi}^{(h)}$ have to be keep in mind, and using them as multipliers in (2).

The Restrictions (a)–(c). The system of restrictions (conditions (a)–(c), point no. 2) setting the set of X of admissible solutions for both the main problem (1) and for auxiliary problem (2) registers in the form:

$$\sum_{i=1}^{n} l_i \cdot x_{ri} \leq v_r, r = 1, 2, \ldots, p; \tag{3}$$

$$\sum_{r=1}^{p} x_{ri} = 1, i = 1, 2, \ldots, n; \tag{4}$$

$$\sum_{r=1}^{p} v_r \cdot H_{qr}(x) \leq N_q, q \in Q; \tag{5}$$

$$x_{ri} \in \{0, 1\}, r = 1, 2, \ldots, p; i = 1, 2, \ldots, n \tag{6}$$

where in (5) the value v_r is length of page r, $r = 1, 2, \ldots, p$. The system (3)–(6) contains $p + n + |Q|$ non-trivial correlations. Note that constraints (3)–(5) corresponding to restrictions (a)–(c) respectively, which mentioned above (point no. 2). The function $H_{qr}(x) : H_{qr}(x) = 1$, if page $S_r \in R(q, x)$ and $H_{qr}(x) = 0$, otherwise, i.e. the function $H_{qr}(x)$ is characteristic function of the $R(q, x)$ set.

Under given q and r it is easy to calculate $H_{qr}(x)$ via elements of the matrix x, namely if $q = (i_1, i_2, \ldots, i_{m(q)}) \in Q$ then $H_{qr}(x) = \max_{1 \leq j \leq m(q)} x_{rij}$.

Here it worth to remember that the matrix $x = (x_{ri})_{p \times n}$ has the only ones positioned in any column including columns $i_1, i_2, \ldots, i_{m(q)}$, see correlation (4) above.

In conclusion of the point let us pay attention at steps which will be useful under solution of auxiliary problem (2) with functional $F^{(h)}(x)$:

0. *Step.* It has to be given: an initial matrix $= (x_{ri})_{p \times n}$. As the matrix x, possible to take the matrix $x^0 = (x_{ri}^0)_{p \times n}$ is using in system before code reorganization; values $l_1, l_2, \ldots, l_n$ (lengths of blocks) and values $v_1, v_2, \ldots, v_p$ (lengths of pages) and values k, h are considered to be given. The value k-parameter of WS(k, t) strategy and the frame dimension too (see Fig. 2, point no. 3, 3.1) is considered to be given. The value $h \geq 1$ is a number of runs of the code for the functional $F^{(h)}(x)$ to be formed.

1. *Step.* Integer parameter $\lambda \geq 1$ is considered to be given. The value $\lambda \geq 1$ is parameter of a common procedure which could be chosen by anybody who is interested in to determine the set Q. It means that if after λ additional runs of the code with a try to find new c.s., the set Q has not changed, then it is considered, that the set Q is found. Let the set of c.s. Q according the descriptions in the sections no. 3.1, 3.3 and notice above is determined (see also Fig. 3).

2. *Step.* Integer parameter $N > 0$ considered to be given The system constant N, which very likely known in advance, will be used instead of constant N_q in restriction (5) above. Here, the constant N has to limited a dimension of any working set $R(q, x)$, then in (5) we put $N_q = N$ for any $q \in Q$. An option to put $N = \max_{q \in Q} N_q$ implies to know any constant $N_q, q \in Q$ and it is possible to do, under not large cardinality of Q and extra calculation. However under the real cardinality of Q, which we may come across, the choice of $N = \max_{q \in Q} N_q$ not acceptable and not appropriate for us.

3. *Step.* Calculation of coefficients $E_{qi}^{(h)}$. Remember, the number h is a number of runs to calculate coefficients $E_{qi}^{(h)}$ in (2). Here we can use convenient for us the formula:

$$E_{qi}^{(j)} := \frac{k-1}{k} \cdot E_{qi}^{(j-1)} + \frac{1}{k} \cdot \xi_{qi}^j, q \in Q, i = 1, 2, \ldots, n; \ i \notin q; \ j = 1, 2, \ldots, h$$

where number j is $j-$ th run of the code. Starting value $E_{qi}^{(0)} = 0$, for all $q \in Q, i \in \{1, 2, \ldots, n\}; i \notin q$. Here after any run, including $j-$ th run, the values $\xi_{qi}^j, q \in Q, i = 1, 2, \ldots, n; i \notin q; j = 1, 2, \ldots, h$ are available. Indeed, during any run of the code, say it be $j-$ th run, the reference string to blocks, by and by, becaming available and not difficult to calculate values ξ_{qi}^j. Namely, it is possible, to calculate them, while the frame moving along axis t from left to the right (see Fig. 2, point no. 3), $q \in Q, i = 1, 2, \ldots, n; i \notin q; j = 1, 2, \ldots, h$. Here, it worth to introduce, for any ξ_{qi}^j wanted to be calculate, the special counter, let it be η_{qi}^{jt}, where for moment $t = 1, 2, \ldots, \gamma$, under the frame movement along axis t (see Fig. 2, point no. 3), the value

$$\eta_{qi}^{jt} = \begin{cases} \eta_{qi}^{j(t-1)} + 1, & \text{if current c.s. is } q \text{ and the next reference to block is } i \\ \eta_{qi}^{j(t-1)}, & \text{otherwise} \end{cases}$$

and after the current run finished, it is necessary to put $\xi_{qi}^j = \eta_{qi}^{j\gamma}$, here γ is last moment of $j-$ th run (see Fig. 2, point no. 3). Starting values $\xi_{qi}^j = 0, \eta_{qi}^{j0} = 0$ for any $q \in Q, i = 1, 2, \ldots, n; i \notin q$. In general, here not necessary to introduce array of variables: $\eta_{qi}^{j0}, \eta_{qi}^{j1}, \ldots, \eta_{qi}^{j\gamma}$. In programmer's style, it is enough to use only one variable, for example, η_{qi}^j and rewrite formula above: $\eta_{qi}^j := \eta_{qi}^j + 1$, if current c.s. is q and the next reference to block is i, otherwise $\eta_{qi}^j := \eta_{qi}^j$. Here the random sequence of the initial data from $D = \{\theta\}$ in order to do runs, and corresponding reference strings to the blocks after any run is considered to be available.

4. *Step.* Solution of problem (2) with functional $F^{(h)}(x)$ with restrictions (3)–(6). The question is reduced, what kind algorithm will be acceptable for solution problem (2). Here, it is possible try to use ideas imbedded into classical algorithms, starting from implicit enumeration algorithm; branch and bound algorithm; pseudo Boolean programming method; and other algorithms.

5 Conclusion

This paper has constructed a combinatorial analytical model for restructuring ill-structured programs (code) in page systems of virtual memory using the Working Set (WS) swapping strategy. Despite its practical applications, the problem has substantial theoretical interest, given that some versions of the problem belong to the NP class.

The approach we've developed is rooted in a key factor: the existence of restructuring invariants, or elements of set Q, which are entirely determined by the WS swapping strategy. Each invariant or control state $q \in Q$ corresponds to a respective node of the Boolean. We present a geometric interpretation of the computational process as a random walk over these Boolean nodes. This visualization simplifies understanding of the process and geometrically explains the construction of the main and auxiliary problem functionals, as well as the estimation functional for both. The estimation functional plays a crucial role in constructing an optimization algorithm, which will be the next step of this research. We have discussed some details of such an algorithm. Additionally, the geometric approach provides a basis to reduce the dimension of the combinatorial problem corresponding to the model presented in this paper.

Earlier developed approaches, primarily based on cluster techniques, offer practical solutions with an unknown approximation error to the optimal solution of (1). If the distribution laws of the introduced random variables are known in advance, then the model constructed in this paper provides a basis to find an optimal solution to the main problem (1). Conversely, if the distribution laws of the random variables are unknown, the model offers a means to find the optimal solution for the auxiliary problem (2). Under certain conditions for the initial data, an optimal solution of (2) will be an ε-optimal solution for the main problem (1).

Our results focus on the combinatorial aspect of restructuring, differing from cluster or graph approaches. We believe these results will prove useful for similar research, such as SVM-like multi-computer systems.

It's worth noting that to date, the WS strategy has primarily been used as a theoretical basis for research for comparison or ancillary purposes. It's often considered too expensive for practical implementation. However, in our case, if a program (code) has a block structure, our findings suggest the possibility of constructing a fast and inexpensive WS-like swapping algorithm, amplifying the value of the results obtained.

References

1. Ngetich, M.K.Y., Otieno, C., Kimwele, M.: A model for code restructuring, a tool for improving systems quality in compliance with object oriented coding practice. IJCSI Int. J. Comput. Sci. Issues **16**(3), pp. 32–36 (2019)

2. Marian, Z., Czibula, I.-G., Czibula, G.: A hierarchical clustering-based approach for software restructuring at the package level. In: 2017 19th International Symposium on Symbolic and Numeric Algorithms for Scientific Computing (SYNASC), pp. 239–246 (2017)
3. Ferrari, D.: Computer Systems Performance Evaluation, 1st edn. Prentice Hall, 1 April 1978. 554 p.
4. Machado, J.P.L., Paula-Sobrinho, E.V.P., Maia, M.A.: Anti-bloater class restructuring: an exploratory study. J. Softw. Evol. Process **34** (2022)
5. Kaur, S., Kaur, A., Dhiman, G.: Deep analysis of quality of primary studies on assessing the impact of refactoring on software quality. Mater. Today Proc., January 2021
6. Masuda, T., Shiota, H., Noguchi, K., Ohki, T.: Optimization of program organization by cluster analysis. In: Proceedings of the IFIP Congress, pp. 261–266 (1974)
7. Dyusembaev, A.E.: Correct models of program segmenting. J. Pattern Recognit. Image Anal. **3**(6), 187–204 (1993)
8. Dyusembaev, A.E.: Mathematical Models of Program Segmentation. M: Fizmatlit (Nauka, MAIK) (2001), 208 p.
9. Foulds, L.R.: Combinatorial Optimization. Springer, Heidelberg (1984). 280 p.
10. Kaur, S., Singh, P.: How does object-oriented code refactoring influence software quality? Research landscape and challenges. J. Syst. Softw. **157**, 110394 (2019)
11. Tenorio, D., Bibiano, A.C., Garcia, A.: On the customization of batch refactoring. In: IEEE/ACM 3rd International Workshop on Refactoring (IWoR) (2019). https://doi.org/10.1109/IWoR.2019.00010
12. Harris, M.: Unified Memory for CUDA Beginners, 19 June 2017. https://developer.nvidia.com/blog/unified-memory-cuda-beginners/

Distributed Systems Management

Probabilistic Resources Allocation with Group Dependencies in Distributed Computing

Victor Toporkov$^{(\boxtimes)}$, Dmitry Yemelyanov, and Artem Bulkhak

National Research University "MPEI", Moscow, Russia
{ToporkovVV, YemelyanovDM, BulkhakAN}@mpei.ru

Abstract. In this work, we introduce and study a set of tree-based algorithms for resources allocation considering group dependencies between their parameters. Real world distributed and high-performance computing systems often operate under conditions of the resources availability uncertainty caused by uncertainties of jobs execution, inaccuracies in runtime predictions and other global and local utilization events. In this way we can observe an availability over time function for each resource and use it as a scheduling parameter. As a single parallel job usually occupies a set of resources, they shape groups with common probabilities of usage and release events. The novelty of the proposed approach is an efficient algorithm considering groupings of resources by the common availability probability for the resources' co-allocation. The proposed algorithm combines dynamic programming and greedy methods for the probability-based multiplicative knapsack problem with a tree-based branch and bounds approach. Simulation results and analysis are provided to compare different approaches, including greedy and brute force solution.

Keywords: Distributed Computing · Resource · Uncertainty · Availability · Probability · Job · Group · Knapsack · Branch and Bounds

1 Introduction and Related Works

High-performance distributed computing systems, such as Grids, cloud, and hybrid infrastructures, provide access to large amounts of resources. These resources are typically required to execute parallel jobs submitted by users and include computing nodes, data storages, network channels, software, etc. The actual requirements for resources amount and types needed to execute a job are defined in resource requests and specifications provided by users [1–5]. Distributed computing systems organization and support bring certain economical expenses: purchase and installation of machinery equipment, power supplies, user support, etc. As a rule, users and service providers interact in economic terms and the resources are provided for a certain payment. Economic models [3–5] are used to efficiently solve resource management and job-flow scheduling problems in distributed environments such as cloud computing and utility Grids. Majority of scheduling solutions for distributed environments implement scheduling strategies on a basis of efficiency criteria [1–5].

V. Malyshkin (Ed.): PaCT 2023, LNCS 14098, pp. 151–165, 2023.
https://doi.org/10.1007/978-3-031-41673-6_12

Traditional models consider scheduling problem in a deterministic way. Such an approach is sometimes justified by the strict market rules for resources acquisition and utilization during the purchased period of time. Commercial Grids and cloud service providers usually own full control over the resources and may reliably consider their local schedules for some scheduling horizon time [1, 3]. Besides, market-based interactions and QoS constraints compliance require deterministic model for the resources utilization profile. Thus, it is convenient to represent available resources as a set of slots: time intervals when particular nodes are idle and may be used for user jobs execution [4–8]. However general distributed computing systems with non-dedicated resources usually cannot rely on deterministic utilization schedules and instead make predictions based on the utilization predictions and probabilities [9–12]. The probabilities of the resources' availability and utilization at any given time may originate from jobs execution and completion time uncertainties, local activities of the resource provider, maintenance, or numerous failure events. Particular utilization characteristics and patterns usually strongly depend on the resource types. However, according to [9] about 20% of Grid computational nodes exhibit truly random availability intervals.

The scheduling problem in Grid is *NP*-hard due to its combinatorial nature and many heuristic solutions have been proposed. When scheduling under uncertainties, proactive and reactive approaches are usually distinguished [12]. Proactive algorithms concentrate on the resources' utilization predictions and heuristic-based advanced resources allocations and reservations. Reactive algorithms analyze current state of the computing environment and make decisions for jobs migration and rescheduling. Both types of algorithms may be used in a single system to achieve even greater resource usage efficiency. The resources availability predictions for the considered scheduling interval may be obtained based on the historical data processing, linear regression models or with help of expert and machine learning systems [9–11]. In [10], a set of availability states is defined to model resource behavior and probabilities state transitions. On the other hand, sometimes it is possible to identify distributions of resources utilization and availability intervals [9]. Economic scheduling models are implemented in modern distributed and cloud computing simulators GridSim and CloudSim [13]. They provide reliable tools for resources co-allocation but consider price constraints on individual nodes and not on a total window allocation cost. However, as we showed in [6], algorithms with a total cost constraint are able to perform the search among a wider set of resources and increase the overall scheduling efficiency. Algorithms [14–16] implement knapsack-based slot selection optimization according to a probability-based criterion with a total job execution cost constraint.

This paper extends scheduling algorithms and model presented in [14–16]. We propose proactive algorithms for resources selection and co-allocation computing environments with non-dedicated resources and corresponding availability uncertainties. The uncertainties are modeled as resources availability events and probabilities: a natural way of machine learning and statistical predictions representation [16]. Common resources' allocation and release times are modeled with interdependent resource groups.

The novelty of the proposed approach consists in a dynamic programming scheme performing resources selection with a total availability criterion maximization. The paper is organized as follows. Section 2 presents availability-based scheduling problem and several greedy, knapsack and branch and bounds-based approaches for its solution. Section 3 contains an experiment setup and simulation results obtained for the considered algorithms. Section 4 summarizes the paper and highlights further research topics.

2 Resource Selection Algorithm

2.1 Probabilistic Model for Resource Utilization

In our model we consider a set R of heterogeneous computing nodes with price c_i characteristics under utilization uncertainties. The probabilities (predictions) $p_i(t)$ of the resources' availability and utilization for the whole scheduling interval L are provided as input data. We model a resource utilization schedule as an ordered list of utilization events, such as resource's *allocation, occupation (execution)* and *release* events. An individual job execution on a single resource is modeled as a sequence of *allocation, occupation* (actual execution) and *release* events (see Fig. 1). Additionally, global resources utilization uncertainties, such as maintenance works or network failures, are modeled as a continuous *occupation* events with $P_o \ll 1$ during the whole considered scheduling interval.

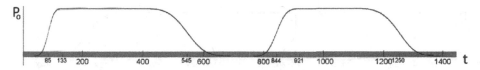

Fig. 1. Example of a single resource occupation probability schedule.

Figure 1 shows an example of a single resource occupation probability P_o schedule. With two jobs already assigned to the resource, there are two resources allocation events (with expected times of allocation at 85 and 844 time units), two resources occupation events (starting at 133 and 921 time units) and two resources release events (expected release times are 545 and 1250 time units respectively). Gray translucent bar at the bottom of the Fig. 1 represents a sum of global utilization events with a total resource occupation probability $P_o = 0.06$. During the whole *execution* interval, the resource's occupation (utilization) probability is assumed as $P_o = 1$. Utilization probability for *allocation* events is modeled by random variable with a normal distribution, and for *release* events - with a *lognormal distribution* to take into account the long tails [15]. Expected allocation and release times are derived from the job's replication and execution time estimations. Corresponding standard deviations depend on the job's features and may be predicted based on user estimations or historical data [9–11, 15]. Hence, in Fig. 1 the resource occupation probability at expected times of allocation and release events are: $P_o(85) = P_o(545) = P_o(844) = P_o(1250) = 0.5$.

However, to execute a job, a resource should be allocated for a specified time period T. Based on the model above, we propose the following procedure to calculate a total availability probability P_a of a resource r during time interval T. P_a describes probability, that the resource r will be fully available and will not be interrupted during T.

1. Retrieve a set of independent utilization events e_i active for the resource r during the time interval T. When a subset of dependent events is active during the interval, then only a single event providing the maximum occupation probability P_o is retrieved. For example, from the *allocation-occupation (execution)-release* events chain only the *execution* event is retrieved with $P_o = 1$.
2. For each independent event e_i a maximum occupation probability during the interval l is calculated: $P_o^{max}(e_i) = \max_{t \in T} P_o(e_i, t)$. Corresponding partial availability probability $P_a(e_i)$ is calculated for each event e_i as a probability that the resource will not be occupied by the event during the interval T: $P_a(e_i) = 1 - P_o^{max}(e_i)$.
3. The resource will be available during the whole-time interval T only in case it will not be occupied by any of the active utilization events. Thus, the total availability probability for the resource r is a product of all partial availability probabilities calculated for independent events e_i:

$$P_a^r = \prod_i P_a(e_i) \tag{1}$$

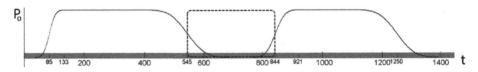

Fig. 2. Example of a resource occupation probability schedule.

For example, consider a resource availability probability for an interval T : [545; 844] presented as a dotted rectangle in Fig. 2. Three independent events are active during the interval: 1) resource release event e_1 with the expected release time at 545 time units, 2) resource allocation event e_2 with the expected allocation time at 844 time units, and 3) a global utilization event e_3 with a constant occupation probability $P_o = 0.06$ (related details were provided with a Fig. 1 example). Corresponding partial occupation and availability probabilities are: $P_o^{max}(e_1) = 0.5$, $P_o^{max}(e_2) = 0.5$, $P_o^{max}(e_3) = 0.06$, while $P_a(e_1) = 0.5$, $P_a(e_2) = 0.5$, $P_a(e_3) = 0.94$. So, the total probability of the resource availability during the whole interval T is $P_a^r = 0.235$.

2.2 Parallel Job Scheduling and Group Dependencies

To execute a parallel job, a set of simultaneously available nodes (a *window*) should be allocated ensuring user requirements from the resource request. The resource request usually specifies number n of nodes required simultaneously for a time period T and a maximum available resources allocation budget C. The total cost of a window allocation is calculated as $C_W = \sum_{i=1}^{n} T * c_i$, where c_i is resource i price for a single time unit.

These parameters constitute a formal generalization for resource requests common among distributed computing systems and simulators [13–16]. Period T of the resources acquisition is usually the same for all resources selected for a parallel job. Common allocation and release times ensure the possibility of inter-node communications during the whole job execution. In this way, the *total window availability* is a function of availability probabilities of all the selected resources during the considered time interval T. More formally, when a set of n resources is selected for a job, the total window availability P_a^w during the expected job execution interval can be estimated as a product of availability probabilities $P_a^{r_i}$ of each *independent* window nodes:

$$P_a^w = \prod_{i}^{n} P_a^{r_i}. \tag{2}$$

Here $P_a^{r_i}$ can be calculated for each resource by the algorithm described in Sect. 2.1. If any of the window nodes will be occupied during the expected job execution interval (i.e., $P_a^{r_i} = 0$), the whole parallel job will be postponed or even aborted. Therefore, in general, the window allocation procedure should consider *maximization of the total probability of availability* $P_a^w \rightarrow$ max. Based on the model above the general statement of the window allocation problem is as follows: during a scheduling interval L allocate a subset of n nodes with performance $p_i \geq p$ for a time T, with common allocation and release times and a restriction C on the total allocation cost. As a target optimization criterion, we assume maximization of the whole window availability probability (2).

As we additionally showed in [14, 15], this *general problem can be reduced to the following task*: at a given time t, which defines the set and state of m available resources, allocate a subset of n nodes with a restriction C on their total allocation cost while performing maximization of their total availability probability (2). In [14, 15] we proposed several approaches to solve the problems above. However, their statement and solution assume *independence* of individual resources as well as their utilization events. That is why in (2) we calculate the total window availability as a product of the availability probabilities of its elements.

In a more general and realistic model, the resources and their utilization events *are not independent*. On the contrary, there are group dependencies between the resources' parameters. The most typical example of such a dependency is a result of a parallel job execution. When a parallel job is scheduled, a set of selected resources is allocated for a common period T. That is, all the selected resources will share allocation, occupation, and release times. So, they should be modeled with a common chain of *allocation-occupation-release* events. In another words, these resources have a *group dependency*.

Figure 3 shows example of utilization events modeled for a parallel job, which requested three nodes. Red areas present resources' utilization probability for allocation and release events. As the exact allocation and release times are unknown, the corresponding occupation probabilities $P_o(t) < 1$. Green areas show execution event with the occupation probability $P_o = 1$. The main issue is that criterion (2) becomes inaccurate when applied to a resources' set with many internal group dependencies. For example, in Fig. 3 if we consider total availability probability of resources 1, 4 and 5 at time

$t = 400$, criterion (2) will calculate it as a product $P_a^w = P_a^1 * P_a^3 * P_a^4$. However, as these resources are used by the same parallel job (and have a common group dependency), their actual total availability probability $P_a^w = P_a^1 = P_a^3 = P_a^4 \geq P_a^1 * P_a^3 * P_a^4$.

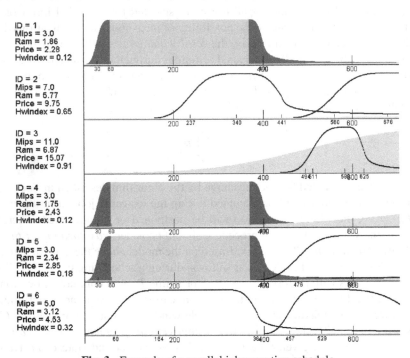

Fig. 3. Example of a parallel job execution schedule.

To describe it more formally we consider a set of groups G over the set R of the available resources. Each component group $G_i \in G$ represents a subset of resources $r_j \in R$ with a common group dependency. For example, one scheduled job, like in the example above, forms a single group G_i which includes all the resources selected for the job. So, for example, if one resource $r_j \in G_i$ is selected for a window W, the common group availability $P_a^{G_i}$ should be used for calculation of a total W availability probability P_a^w. However, additionally selecting any other resources from G_i will not affect P_a^w, as their group probability component $P_a^{G_i}$ is already considered.

So, the total window W availability probability can be calculated as follows:

$$P_a^w = \prod_i^{n^*} P_a^{G_i}, \qquad (3)$$

where n^* is a number of different groups used for the window W, and $P_a^{G_i}$ is availability probability for each different group G_i used for the window. Group G_i is added to (3) if at least one of its resources is selected for the window. It is worth noting, that in the extreme case each group G_i can contain only one resource, and thus (3) will converge to

(2). In this paper we propose and study resources allocation algorithm which performs (3) $P_a^w \rightarrow$ max optimization considering economic constraint on the total window cost and group dependencies G. However firstly we should introduce helper algorithms performing (2) $P_a^w \rightarrow$ max optimization without the group dependencies configuration.

2.3 Direct Solutions of the Resources Allocation Problem

Let us discuss in more details an algorithm which allocates an optimal (according to the probability criterion P_a^w) subset of n resources from the set R of m available resources with a limit C on their total cost.

Firstly, we consider maximizing the following total resources availability criterion $P_a^w = \prod_j^n p_a^{r_j}$, where $p_a^{r_j} = p_j$ is an availability probability of a single resource $r_j \in R$ during a considered interval T. In this way we can state the following problem of an n - size window subset allocation out of m nodes:

$$P_a^w = \prod_j^m x_j p_a^{r_j} \rightarrow \max, \sum_j^m x_j c_j \le C, x_j \in \{0, 1\}, j = 1..m, \sum_j^m x_j = n, \quad (4)$$

where c_j is total cost required to allocate resource r_j, x_j - is a decision variable determining whether to allocate resource r_j ($x_j = 1$) or not ($x_j = 0$) for the current window.

This problem relates to the class of integer linear programming problems, and we used 0–1 knapsack problem as a base for our implementation. The classical 0–1 knapsack problem with a total weight C and items-resources with weights c_j and values p_j have a similar formal model except for extra restriction on the number of items required: $x_1 + x_2 + \cdots + x_m = n$. Therefore, we implemented the following dynamic programming recurrent scheme:

$$f_j(c, v) = \max\{f_{j-1}(c, v), f_{j-1}(c - c_j, v - 1) * p_j\},$$
$$j = 1, .., m, c = 1, .., C, v = 1, .., n, \quad (5)$$

where $f_j(c, v)$ defines the maximum availability probability value for a v-size window allocated from the first j resources of m for a budget c. After the forward induction procedure (4) is finished the maximum availability value $P_{a\,max}^w = f_m(C, n)$. x_j values are then obtained by a backward induction procedure.

An estimated computational complexity of the presented knapsack-based algorithm *KnapsackP* is $O(m * n * C)$.

Another approach for n-size window allocation is to use a more computationally efficient greedy approach. We outline four main greedy algorithms to solve the problem (3). The task is to select n out of m resources providing maximum total availability probability P_a^w with a constraint on their total allocation cost n.

1. *MaxP* selects first n nodes providing maximum availability probability p_j values. This algorithm does not take into account total usage cost limit and may provide infeasible solutions. Nevertheless, *MaxP* can be used to determine the best possible availability options and estimate a budget required to obtain them.
2. An opposite approach *MinC* selects first n nodes providing minimum usage cost c_j or an empty list in case of exceeding a total cost limit C. In this way, *MinC* does not perform any availability optimization, but always provides feasible solutions when it

is possible. Besides, *MinC* outlines a lower bound on a budget required to obtain a feasible solution.

3. Third option is to use a weight function to regularize nodes in an appropriate manner. *MaxP/C* uses $w_j = p_j/c_j$ as a weight function and selects first n nodes providing maximum w_j values. Such an approach does not guarantee feasible solution, but nonetheless performs some availability optimization by implementing a compromise solution between *MaxP* and *MaxC*.

4. Finally, we consider a composite approach *GreedyUnited* for an efficient greedy-based resources allocation. The algorithm consists of three stages.

   a. Obtain *MaxP* solution and return it if the constraint on a total usage cost is met.
   b. Else, obtain *MaxP/C* solution and return it if the constraint on a total usage cost is met.
   c. Else, obtain *MinC* solution and return it if the constraint on a total usage cost is met.

This combined algorithm *GreedyUnited* is designed to perform the best possible greedy optimization taking into account a restriction on a total resources' allocation cost C.

Estimated computational complexity for the greedy resources' allocation step is $O(m * \log m)$. More details regarding the algorithms above are provided in [14–16].

2.4 Resources Allocation Algorithms with Group Dependencies

Based on *KnapsackP* and *GreedyUnited* implementations above we propose the following algorithm for a general resource allocation problem considering group dependencies between the available resources. It takes as input set R of the available resources (each resource is characterized with cost c_i) and set G of groups over R (each group G_i has a common availability probability p_i). The algorithm then allocates a subset of n resources with a restriction C on their total cost while performing maximization of their total availability probability (3). The problem is solved by branch and bounds method by maintaining max-heap data structure H containing interim candidate solutions S_j. The higher the achieved availability probability P_a^w (3) or its upper bound, the closer the solution S to the top of the heap H. For each solution S we maintain two subsets of groups that should (G^+) and should not (G^-) be used in the current solution. Both G^+ and G^- are initialized as empty sets. Additionally, we consider subset G^0 as all groups from G not included in G^+ or G^-, so G^0 is initialized as G.

Initial candidate solution S^0 with empty $G^0 = G$ and empty sets G^+ and G^-, is placed into H with $P_a^w = -infinity$. Next, we perform the following steps.

1. Retrieve next solution candidate S from H. If S is marked as valid solution, then return S as a result, end of the algorithm.
2. Prepare list of resources R_s to calculate P_a^w for S.
   a. Init R_s as empty set.
   b. For each group G_j from G^+ add the cheapest resource to the solution window W_s with the $p_i = P_a^{G_j}$; add other resources from this group $r_i \in G_j$ to R_s with $p_i = 1$.

c. For each group G_j from G^0 add all resources $r_i \in G_j$ to R_s with $p_i = \sqrt[k]{P_a^{G_j}}$, where k is number of resources in G_j.

3. Use algorithm *KnapsackP* or *GreedyUnited* to perform direct solution of S to allocate resources into W_s (it can be partially filled during step 2.b) from set R_s of prepared resources with (2) $P_a^w \rightarrow$ max optimization without group dependencies.

4. Check if the resulting solution is valid.

a. If all resources from W_s are included in groups from G^+, then put this solution S into H with key P_a^w and mark it as a *valid solution*.

b. If at least one resource r_s from W_s is included in some group G_s from G^0, then we need to split this solution S into two candidates: S^+ and S^-. For S^+ remove group G_s from G^0 and add into G^+. For S^- remove group G_s from G^0 and add into G^-. Put both solution candidates S^+ and S^- into H with key P_a^w as an upper estimate.

5. Go to step 1.

The algorithm above performs branch and bounds approach by splitting candidate solutions by sets of resources groups G^+ and G^- required to use or skip correspondingly. A special resource set R_s preparation in step 2 allows us to use (2) optimization algorithms and obtain either a final valid solution or a candidate solution with pretty accurate upper estimate. The algorithm finishes when the next solution obtained from the max-heap data structure is a valid solution composed of resources from G^+ groups and, thus, its P_a^w calculated with (2) satisfies rules for group dependencies availability calculations (3).

3 Simulation Study

3.1 Considered Algorithm Implementation

For the simulation study we consider and compare the following algorithm implementations.

1. Firstly, we implemented *brute-force* algorithm to solve the resources allocation problem with (3) $P_a^w \rightarrow$ max optimization. We used this algorithm for a preliminary analysis in small experiments with up to 21 resources to compare its optimization efficiency with other approaches.

2. Next, we prepared three implementations of a general branch and bounds algorithm described in Sect. 2.4. First implementation *KnapsackGroup* uses *KnapsackP* for all interim allocations during the algorithm step 3. *Greedy* performs interim optimizations at step 3 with *GreedyUnited* algorithm. Finally, *Greedy+* runs *GreedyUnited* for all interim optimizations, but once the solution is found, the final solution optimization is performed again using more accurate *KnapsackP* approach.

3. Finally, we consider *KnapsackP (KnapsackSingle)* as standalone algorithms for the comparison. This algorithm does not support group dependencies and performs (2) $P_a^w \rightarrow$ max optimization. The obtained solution is then recalculated accordingly to (3) to compare it to the algorithms above.

For the simulation study we execute and collect resulting data for all the considered algorithms (*BruteForce, KnapsackGroup, Greedy, Greedy+* and *KnapsackSingle*) in different resource environments with randomized characteristics c_i, p_i and group dependencies. An experiment was prepared using a custom distributed environment simulator [6, 14, 15]. For our purpose, it implements a heterogeneous resource domain model: nodes have different usage costs and availability probabilities (https://github.com/dmi eter/proba-sch/commits/master). Each node supports a list of active global and local job utilization events. Figure 3 shows an example of such an environment with many resources and a Gantt chart of the utilization events.

Additionally, we generate random uniformly distributed group dependencies between the resources. So, the resources allocation problem can be defined with the following parameters: N – number of available resources (each characterized with cost c_i and availability probability p_i), G – number of different groups (containing random non-intersecting subsets of resources), n – number of resources required for allocation and C – available budget, i.e. constraint on the total cost of the selected resources.

3.2 Proof of Optimization Efficiency

The first experiment series studies algorithms optimization and computational efficiency in comparison with *BruteForce* approach. Brute force is usually inapplicable in real-world tasks due to its exponential computational complexity. However, it guarantees exact optimization solution, and can be used to evaluate optimization characteristics of other considered algorithms. During each simulation experiment, the resources allocation was independently performed by algorithms *BruteForce, KnapsackGroup, Greedy, Greedy+* and *KnapsackSingle*. The comparison is obtained with different values of G, n, C of the allocation problem. As *BruteForce* applicability is limited, firstly we performed resources allocation simulation with only $N = 21$ available resources.

Figure 4 shows resulting availability probability P_a^w depending on number $n \in [1; 21]$ of requested resources in environment with $N = 21$ available resources, $G = 8$ different groups and without the total cost restriction ($C = \sum_i^N c_i$). The main result is that proposed algorithms *KnapsackGroup, Greedy* and *Greedy+* provided the same P_a^w value as *BruteForce* (that is why they are not presented in Fig. 4). *KnapsackGroup* theoretically guarantees exact problem solution in integers and is expected to provide results identical to *BruteForce*. Greedy algorithms provided optimal solution due to the lack of the total cost limit (see *GreedyUnited* and *MaxP* descriptions in Sect. 2.3). However, *KnapsackSingle* in most cases failed to provide optimal solution with up to 5% lower availability probability compared to *BruteForce*. The equality is achieved only in two simplified scenarios with $n = 1$ and $n = 21$, when group dependencies are not relevant for the problem.

Figure 5 shows actual algorithms' execution time required to achieve allocation results from Fig. 4. As can be seen, *BruteForce* calculation time dramatically increases for $n \in [7; 15]$ and exceeds half a second for $n = 11$. This is explained by the combinatorial nature of selecting subset of n from N available resources. Even the most computationally complex *KnapsackGroup* algorithm, which combines pseudo polynomial 0–1 knapsack implementation with branch and bounds approach is presented in Fig. 5 as a straight line 100 times lower compared to the *BruteForce* maximum. *Greedy* approaches were up to

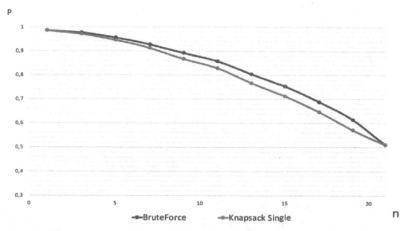

Fig. 4. Simulation results: resulting availability probability P_a^w depending on number n of requested resources.

1000 times faster than *BruteForce*. So, according to the trend in Fig. 5, in environments with $N > 25$ *BruteForce* becomes practically inapplicable and other exact algorithms and approximations should be considered. The accuracy of such approximations in general should be estimated with the economical restriction C on the total window allocation cost.

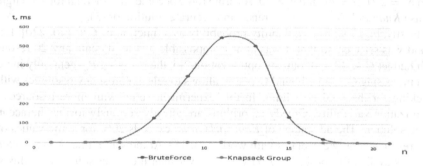

Fig. 5. Simulation results: average calculation time depending on number n of requested resources.

Figure 6 shows how window availability probability depends on the allocation budget $C \in [30; 120]$ in problem setup with $n = 8$, $G = 8$ and $N = 21$. In this environment only *KnapsackGroup* was able to obtain exact solutions (identical to *BruteForce*) for all C values. Additionally, *KnapsackGroup* provides almost constant 5% advantage over *KnapsackSingle*. The results of Greedy algorithms are also within 5% of the exact solution and reaches *BruteForce* for $C > 90$. In general, the obtained simulation result confirms accuracy of *KnapsackGroup* algorithm and gives an approximate estimate of the accuracy of the more computationally simple algorithms.

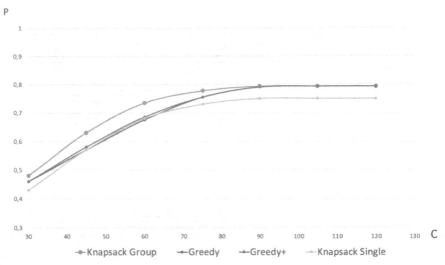

Fig. 6. Simulation results: resulting availability probability P_a^w depending on the budget $C \in$ [30; 120].

3.3 Practical Optimization Efficiency Study

Next experiment series studies proposed algorithms in more complex problem settings with $N = 200$, $G = 40$ and $n = 20$. As brute force becomes impractical for such figures, we use *KnapsackGroup* as a reference and accurate solution of (3).

Firstly, Fig. 7 shows availability probability as a function of $C \in$ [40; 220]. Lower bound was selected so that it was almost impossible just to allocate any 20 resources with budget $C < 40$, without any optimization. So, the resulting P_a^w generally increase with increasing C. Upper bound $C > 200$ allows to select almost any resources without checking for the total cost limit. In this experiment setup with more resources and optimization variability, Greedy algorithms are already seriously losing the accuracy of the solution. The advantage of *KnapsackGroup* exceeds 20% for some values of C. This result generally correlates with works [14, 15]. At the same time, *KnapsackSingle* provides availability probability only 10% lower than the exact solution. In this way, the absence of group dependencies information turns out advantageous compared to the accuracy of greedy approximations of the multiplicative knapsack problem. Only in scenarios with $C > 200$, i.e., without the cost restriction, *Greedy* is able to outperform *KnapsackSingle* in environment with group dependencies between the resources.

Another important factor for the practical applicability is the algorithms' calculation time presented in Fig. 8 for the same environment settings. The obvious trend is that tighter restrictions on the budget C cause a strong increase in working time for branch and bounds - based algorithms (*KnapsackGroup*, *Greedy*, *Greedy+*). This is explained by the necessity to select resources with respect to the C constraint, rather than by the target criterion.

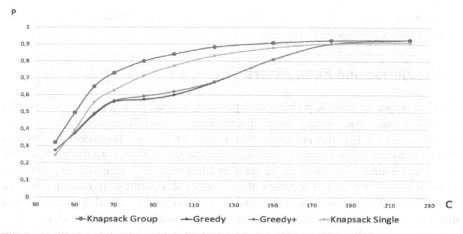

Fig. 7. Simulation results: resulting availability probability P_a^w depending on the budget $C \in$ [40; 220].

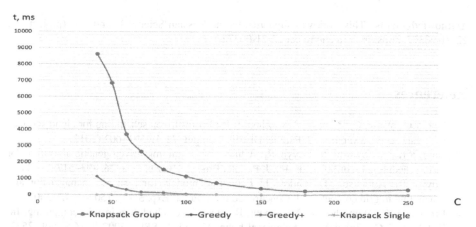

Fig. 8. Simulation results: average calculation time depending on the budget C.

And this strategy requires consideration of more different groups and splitting in branch and bounds approach. For example, with $C = 40$, an average size of the solution tree for *KnapsackGroup* was almost 5000 elements causing nearly 8 s of the execution time. And with $C = 120$ the tree size decreased to nearly 100 elements leading to a sub second execution time. Similar calculation time trend applies to Greedy tree algorithms as well.

Thus, based on Figs. 7, 8 we conclude, that with tight economical budget restrictions the most practically adequate option is a simple multiplicative 0–1 knapsack algorithm [14], as such problem setup requires greater emphasis on the cost optimization and less on the groups' combinations. With a looser cost restriction ($C \in$ [100; 200] in our experiment) tree-based *KnapsackGroup* becomes a preferred option as it provides exact optimization solution for an adequate calculation time. Finally, when there is no cost restriction, a tree-based *Greedy* algorithm can provide exact optimization result in the least amount of calculation time.

4 Conclusion and Future Work

In this work, we address the problem of dependable resources co-allocation for parallel jobs in distributed computing with group dependencies over the resources. Such group dependencies usually define utilization events common for subsets of resources, such as simultaneous allocation or release events. To handle this problem, we designed several branch and bounds algorithms based on a multiplicative 0–1 knapsack problem.In a simulation study we proved accuracy of the proposed algorithms in comparison with a brute force approach, estimated their calculation time and practical applicability in a more complex scheduling problems with up to 200 available computing nodes.

Future work will concern additional optimization in the algorithms' complexity and calculation time. In addition, we plan to consider similar allocation task based on an additive 0–1 knapsack problem.

Acknowledgments. This work was supported by the Russian Science Foundation (project No. 22-21-00372, https://rscf.ru/en/project/22-21-00372/).

References

1. Lee, Y.C., Wang, C., Zomaya, A.Y., Zhou, B.B.: Profit-driven scheduling for cloud services with data access awareness. J. Parallel Distrib. Comput. **72**(4), 591–602 (2012)
2. Garg, S.K., Konugurthi, P., Buyya, R.: A linear programming-driven genetic algorithm for meta-scheduling on utility grids. Int. J. Parallel Emerg. Distrib. Syst. **26**, 493–517 (2011)
3. Buyya, R., Abramson, D., Giddy, J.: Economic models for resource management and scheduling in grid computing. J. Concurr. Comput. Pract. Exp. **5**(14), 1507–1542 (2002)
4. Ernemann, C., Hamscher, V., Yahyapour, R.: Economic scheduling in grid computing. In: Feitelson, D.G., Rudolph, L., Schwiegelshohn, U. (eds.) JSSPP 2002. LNCS, vol. 2537, pp. 128–152. Springer, Heidelberg (2002). https://doi.org/10.1007/3-540-36180-4_8
5. Kurowski, K., Nabrzyski, J., Oleksiak, A., Weglarz, J.: Multicriteria aspects of grid re-source management. In: Nabrzyski, J., Schopf, J.M., Weglarz J. (eds.) Grid Resource Management. State of the Art and Future Trends, pp. 271–293. Kluwer Academic Publishers (2003)
6. Toporkov, V., Toporkova, A., Bobchenkov, A., Yemelyanov, D.: Resource selection algorithms for economic scheduling in distributed systems. In: ICCS 2011, 1–3 June 2011, Singapore (2011). Procedia Computer Science. Elsevier, vol. 4. pp. 2267–2276
7. Netto, M.A.S., Buyya, R.: A flexible resource co-allocation model based on advance reservations with rescheduling support. In: Technical Report, GRIDSTR-2007–17, Grid Computing and Distributed Systems Laboratory, The University of Melbourne, Australia, 9 October 2007
8. Jackson, D., Snell, Q., Clement, M.: Core algorithms of the maui scheduler. In: Feitelson, D.G., Rudolph, L. (eds.) JSSPP 2001. LNCS, vol. 2221, pp. 87–102. Springer, Heidelberg (2001). https://doi.org/10.1007/3-540-45540-X_6
9. Javadi, B., Kondo, D., Vincent, J., Anderson, D.: Discovering statistical models of availability in large distributed systems: an empirical study of SETI@home. IEEE Trans. Parallel Distrib. Syst. **22**(11), 1896–1903 (2011)

10. Rood, B., Lewis, M.J.: Grid resource availability prediction-based scheduling and task replication. J. Grid Comput. **7**, 479 (2009)
11. Tchernykh, A., Schwiegelsohn, U., El-ghazali, T., Babenko, M.: Towards understanding uncertainty in cloud computing with risks of confidentiality, integrity, and availability. J. Comput. Sci. **36** (2016)
12. Chaari, T., Chaabane, S., Aissani, N., and Trentesaux, D.: Scheduling under uncertainty: survey and research directions. In: 2014 International Conference on Advanced Logistics and Transport (ICALT), pp. 229–234 (2014)
13. Calheiros, R.N., Ranjan, R., Beloglazov, A., De Rose, C.A.F., Buyya, R.: CloudSim: a toolkit for modeling and simulation of cloud computing environments and evaluation of resource provisioning algorithms. J. Softw. Pract. Exp. **41**(1), 23–50 (2011)
14. Toporkov, V., Yemelyanov, D.: Availability-based resources allocation algorithms in distributed computing. In: Voevodin, V., Sobolev, S. (eds.) RuSCDays 2020. CCIS, vol. 1331, pp. 551–562. Springer, Cham (2020). https://doi.org/10.1007/978-3-030-64616-5_47
15. Toporkov, V., Yemelyanov, D., Grigorenko, M.: Optimization of resources allocation in high performance distributed computing with utilization uncertainty. In: Malyshkin, V. (eds.) PaCT 2021. LNCS, vol. 12942, pp. 325–337. Springer, Cham (2021). https://doi.org/10.1007/978-3-030-86359-3_24
16. Toporkov, V., Yemelyanov, D., Bulkhak, A.: Machine learning-based scheduling and resources allocation in distributed computing. In: Groen, D., de Mulatier, C., Paszynski, M., Krzhizhanovskaya, V.V., Dongarra, J.J., Sloot, P.M.A. (eds.) ICCS 2022. LNCS, vol. 13353, pp. 3–16. Springer, Cham (2022). https://doi.org/10.1007/978-3-031-08760-8_1

Multicriteria Task Distribution Problem for Resource-Saving Data Processing

Anna Klimenko$^{(\boxtimes)}$ and Arseniy Barinov

Institute of IT and Security Technologies, RSUH, Kirovogradskaya Street, 25-2,
Moscow, Russia
anna_klimenko@mail.ru

Abstract. In the current paper a question of the resource-saving tasks distribution is under consideration. The problem of computational resource saving is topical because of the enormous data volumes, which are preprocessed partially by the fog- and edge- network layers. In general, scheduling and resource allocation are modeled via combinatorial optimization problems without consideration of the fact that the computational environment is geographically distributed. The consequence of such distribution is that the tasks assigned to some nodes have to transmit the data through some transit network sections. As the data transmission produces workload and consumes time, which degrade the average residual time of the nodes, in this paper we propose the novel problem model, which is structural-parametric and focuses not only on the functional tasks assignment to the nodes, but to the data transmission workload, which disseminates through the data transmission routes. The generic solution method is proposed on the base of multiplicative convolution and random search. The produced results show the positive effect of the workload distribution on the nodes reliability function values.

Keywords: Fog Computing · Tasks Distribution · Reliability

1 Introduction

Nowadays distributed computing is almost ubiquitous, including data processing in the dynamic environments, such as fog and edge network layers. As the data volumes to be processed increase, a lot of problems emerge in this field: resource allocation in the lack of time condition [1, 2], information exchange between the participants of the computational process [3, 4], load balancing [5, 6], energy consumption optimization [7], etc.

Yet, very seldom publications are met, which pay attention to the average residual life of the computational nodes within the fog- and edge- network tiers [8–10]. One can see that the increase of the average residual life of the node, as well as the increase of the reliability function values and the decrease of the failure rate allow to prolong the time of resource expedient exploitation. Besides, some researches were made in this field [11], it has been shown that the reliability function, as well as the average residual life of the nodes depend on the way of computational tasks distribution.

V. Malyshkin (Ed.): PaCT 2023, LNCS 14098, pp. 166–176, 2023.
https://doi.org/10.1007/978-3-031-41673-6_13

However the most part of the papers consider the task distribution process as the distribution within the environment with the fully interconnected nodes, while the situations when there are some transit nodes, which just transmit the information, are out of the scope (Fig. 1).

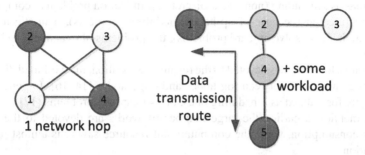

Fig. 1. The comparison between full graph communication environment and the random one.

For example, studies [12, 13] consider the problem of nodes average residual life as a reliability problem and formalize it with the load balancing objective function. Indeed, it was demonstrated that such approach improves the gamma percent time between failures. So the expedient exploitation time of the nodes increases and the exploitation cost, vice versa, decreases. However, the studies considered focus on the fully interconnected communication environment between the nodes, while the need to transmit data through the intermediate network sections is out of the consideration.

Our previous studies present so-called "egoistic"(greedy) technique, which allows to decide, if the computational node is more suitable for the data processing, or it is more profitable for this particular node to transmit this data further (from the data source). This approach can be quite insufficient, because we consider just one node and, actually, this model and strategy can be used in the manner of "greedy" heuristic, which, of cause, can produce the results of very high – or very low quality.

So, the purpose of this research is to create a basic model of tasks distribution problem with the possibility to assign the priorities for the particular nodes. The model must take into account the data transmission processes and the presence of intermediate nodes. We assume the data transmission channels reliable and of the sufficient throughput. As the problem model is produced, the generic technique for its solution must be provided and tested.

The following sections of the paper contain:

– brief overview of the previous works;
– the problem formalization and the generic problem solution technique;
– some experimental results.

2 Previous Work

The field of the tasks distribution through the network environments, including fog and edge, is of a high importance for researchers. It can be mentioned that this field – in the focus of resource allocation – is multidisciplinary, and includes, at least, such problematic areas as scheduling (non-linear discrete optimization problems), computational complexity research (because of np-hard scheduling problems), heuristic and meta-heuristic methods (to solve np-hard problems), the problems of dynamic re-scheduling, and so on.

There are a lot of studies devoted to the resource allocation, the workload distribution between fog and cloud, between fog nodes and edge nodes [14–16]. In general, these problems are formulated as scheduling ones, or as knapsack problems [17].

As is mentioned earlier, the large number of works are devoted to the problem of energy consumption, while the computational resource saving is almost out of the consideration.

Our previous work [18], devoted to the saving of the computational resources, consider only the situation, when the nodes follow the fairway of the "egoistic" behavior of the nodes. Each node considers the question, if it is more profitable for it to process some data or to retransmit it to the next node, which is further from the edge of the network. In these works we consider the basics of the mathematical framework of the subject from the reliability function point of view.

Reliability function value depends on the failure rate of the computational node, while failure rate is connected to the device temperature and workload:

$$\lambda = \lambda_0 \cdot 2^{\frac{\Delta T}{10}} \tag{1}$$

where

- λ is a resulting failure rate,
- λ_0 is the failure rate under conditions of unloaded device,
- ΔT is the temperature difference between the temperature of unloaded device and the temperature of loaded one.

Also, the coefficient can be determined, which connects the node temperature and the workload.

$$\lambda = \lambda_0 \cdot 2^{\frac{kD}{10}} \tag{2}$$

Consequently, the reliability function is determined as follows:

$$P(t) = e^{-\lambda t} = e^{-\lambda t \cdot 2^{kD/10}} \tag{3}$$

where D is the node workload.

Consider the node workload for the node 3 (Fig. 2) as follows (for the case when the node transits some data):

$$D = \frac{w_i}{p_j t_{transfer}} \tag{4}$$

where w_i is the computational complexity of the task, $t_{transfer}$ is the time needed for the data transfer.

Consider the data processing shift discussed in [18], where the data transmission to the cloud takes place (Fig. 2).

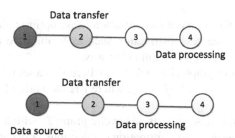

Fig. 2. Data processing shift illustration

With the data processing shift the workload of the processing node is:

$$D = \frac{w_i}{p_j t_{process}} \qquad (5)$$

Obviously, with the data shift $t_{process}$ is bigger than $t_{transfer}$, and so D decreases. And, finally the following "egoistic" rule has been formed in [18].

$$w_{receice} + w_{process} + w_{send} < 2\frac{w_{receive}}{x}, \qquad (6)$$

where x – time fraction for the time of data transfer process of the particular node, $w_{receive}$ – the computational complexity of the data receiving process, w_{send} – the computational complexity of the data transmission process, $w_{process}$ – the computational complexity of the data processing.

So, the data processing shift to the particular node is expedient in terms of reliability when the computational complexity of the data receiving, sending and processing is less than the division of two computational complexities of the data receiving by time fraction of data transfer through the node.

One can see, that according to this rule, the duty of data processing can be passed further and further from the initial point in the network, till the situation, when time constraints will never be met. So, in the study [18] we stated that if the "duty" of tasks processing moves away too far, the nearest node, which is in one network hop distance, is chosen for the data processing.

As it was mentioned, such "egoistic" approach can result in acceptable solution, but there is always risk to get poor results.

So, in the next section, a new problem is formulated – in a way of multiobjective (or multicriteria) optimization problem.

3   Task Distribution Problem for Resource-Saving Data Processing

Previously we consider the "greedy" strategy to distribute the computational tasks through the nodes. In this situation every node doesn't know about the state of its neighbours, takes care just of itself, and estimates just its own computational resource state.

It is obvious, that the problem formalization as an optimization one is much more appropriate: to choose the best distribution such as to improve the overall reliability state. This can be formulated in the following way.

Consider the network graph $G = <V, U>$, where V is a set of computational nodes, U is a set of ribs. $V = \{v_j\} = \{<j, p_j>\}$, where j is the node identifier, p_j is the node performance.

The user operation is described as an acyclic graph T, which vertexes are assigned to tasks, and ribs are assigned to information connections between them:

$T = \{t_i\} = \{<i, w_i, d_i>\}$, where i - is the subtask identifier,
w_i - is the computational complexity of the subtask,
d_i - is the data volume transferred to the network.

The problem solution is the following tasks assignment:

$$A = \begin{vmatrix} t_{ij} & \cdots \\ & \cdots & \\ \cdots & & t_{nm} \end{vmatrix}, \; such \, as \, P_0(\tau) \to \max. \tag{7}$$

where $P_0(\tau)$ is an overall system reliability, t_{ij} are the time moments of task i assignment to the node j.

Consider the objective function. In general, talking of the reliability of the system, we have to consider two basic system structures, with parallel and consequent connection of the elements.

So, parallel connection presupposes that system functions till at least one element functions. Otherwise, consequent connection presupposes that system functions when all elements are alive and performs their tasks. For parallel system $P_0(\tau)$ is described with the following equation:

$$P_0(\tau) = 1 - \prod(1 - P_j(\tau)), P_j(\tau) = e^{-\lambda_j t * 2^{\frac{kD_j}{10}}}. \tag{8}$$

Where $P_0(\tau)$ – is the overall reliability function value for the computational process participants;
D – is the node workload;
k – is the coefficient of node temperature increase depending on the current workload,
t_{ij} – the moment of assignment of task i to the node j.

The consequent system reliability function is described as follows:

$$P_0(\tau) = \prod(P_j(\tau)), P_j(\tau) = e^{-\lambda_j t * 2^{\frac{kD_j}{10}}}. \tag{9}$$

The constraint for this problem is as follows: $\tau \leq t_{decl}$.

Such problem formalization allows to get the resulting tasks distribution, esteeming the overall reliability function for the nodes community. So, the tasks are distributed through the network in terms of "common good" of the community.

Besides, it is obvious that it is hardly possible to present the random network graph in terms of parallel-consequent connections. So, the development of such network model seems to be inexpedient.

To avoid the mentioned issues, we propose to formulate the problem of tasks distribution as the multiobjective optimization problem.

Consider the computational process participants. Following the previous problem statement, we replace the objective function (8–9) with the following: the number of objective functions is equal to the computation process participants, including the data transmission intermediate nodes. So, the objective functions can be represented in the following way:

$$
\begin{aligned}
P_1(\tau) &= e^{-\lambda_0 \tau \cdot 2^{kD_1/10}}; \\
P_2(\tau) &= e^{-\lambda_0 \tau \cdot 2^{kD_2/10}}; \\
&\ldots \\
P_m(\tau) &= e^{-\lambda_0 \tau \cdot 2^{kD_m/10}}.
\end{aligned}
\tag{10}
$$

where m is the number of nodes.

It must be mentioned that with every new tasks assignment, we have some new objective functions in the objective functions vector, due to the distinguishes between the data transmission routes. Data transmission routes emerge because of the need to transfer data between tasks, which can be assigned to the distant nodes.

Then, we use the multiplicative convolution to get one-criterion optimization problem and to solve it. The usage of the multiplicative convolution approach allows to assign some preference weights to the chosen nodes:

$$
P_0(\tau) = \prod_{i=1}^{m} (P_i(\tau))^{\xi_i}
\tag{10}
$$

where ξ_i determines the preference of the particular node, which, for example, we'd like to offload.

The generic method of iterative creating and testing of the tasks assignment effect is presented in the Fig. 3.

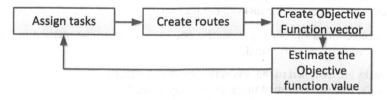

Fig. 3. The generic method of the problem solving

4 Experimental Results

Firstly, the computational problem graph was generated (Fig. 4): the nodes are weighed with the computational complexity of the tasks, the edges are weighed with the data volumes which are transmitted from task to task.

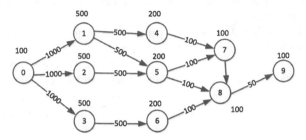

Fig. 4. Computational problem structure

The network graph is taken as is shown in the Fig. 5. We assume the network connections of the sufficient bandwidth, as well as the network is homogeneous.

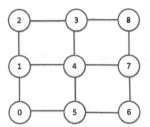

Fig. 5. Network structure

To solve the multiobjective problem described above some software were developed. We used the following general means to search the solution for our tasks distribution problem:

- Simple random search for the solutions improvement;
- Random forming of the routes for those cases when tasks, assigned to the nodes, have to transmit their data through the mediator nodes;
- Random assignment of the tasks to the nodes.

Our problem is a structural-parametric one. According to the method in the Fig. 3 the following steps are performed.

- The tasks are assigned to the nodes in a random manner;
- Then, the routes are formed in a random way as well;
- So we have the community of the nodes, which take part in a computational process, including those ones, which just transmit the data (because the data transmission is a computational task as well);

– The objective function is calculated for the nodes community.

Within the experiment we compare the initial solution (objective function value) and the solution after the random search.

Initial objective function value F = 0,5943 with the tasks distribution shown in the Fig. 6.

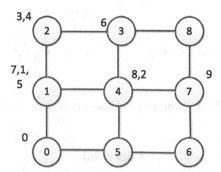

Fig. 6. Initial tasks distribution

After the 10000 search iteration the following objective function value was achieved: F = 0,8777.

The distribution of the tasks is as shown in Fig. 7.

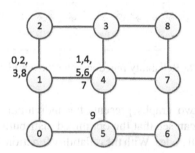

Fig. 7. Final tasks distribution

One can see that despite the increase of the tasks number per node, the number of nodes decreases as well as the number and the length of the data transmission routes, while the latter impacts the time for computational task processing. Also one can see that tasks 0, 2, 3 are assigned to the same node, while the data transfer is quite intensive between those tasks, and task 1 is assigned to the node 4, which is situated within one network hop from the node 1. The tasks with intensive data exchange are assigned to the scattered nodes, the main consequence of this is the need to use routes to transmit data and to waste the time for it.

To complete the experimental research it is useful to estimate the individual reliability functions of the nodes, which are the participants of the computational process (Figs. 8, 9).

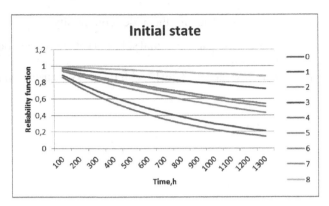

Fig. 8. Reliability functions on the initial stage

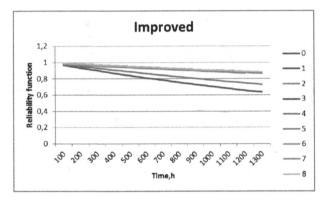

Fig. 9. Reliability functions after the optimization

The comparison of two graphs presents the individual – for each node – effect of the optimization: one can see that the optimized distribution shows better reliability function estimations for all nodes. With the reliability function improvement the potential exploitation time of the nodes increases too.

5   Conclusions

The main contribution of this research is the novel tasks distribution problem formalization with the resource-saving objective function.

The main peculiarity of the model proposed is that we take into account the routes between the nodes with assigned tasks, which affect the nodes set composition, the time of data processing and the final appearance of the objective function vector.

The multiobjective optimization problem solving allows, by means of multiplicative convolution values improvement, to increase the individual nodes reliability function values and so to improve the expedient exploitation time of the nodes.

The experimental results proved the expediency of the approach proposed.

References

1. D'Amato, A., Andrade, W.: A resource allocation model driven through QoC for distributed systems (2022). https://doi.org/10.5772/intechopen.106458
2. Zhijun, G., Gang, C.: Distributed dynamic event-triggered and practical predefined-time resource allocation in cyber–physical systems. Automatica **142**, 110390 (2022). https://doi.org/10.1016/j.automatica.2022.110390
3. Junbin, L., Jie, Z., Victor, L., Xu, W.: Distributed information exchange with low latency for decision making in vehicular fog computing. IEEE Internet Things J. **9**, 18166–18181 (2022). https://doi.org/10.1109/JIOT.2021.3075516
4. Kanika, S., Bernard, B., Jennings, B.: Graph-based heuristic solution for placing distributed video processing applications on moving vehicle clusters. IEEE Trans. Netw. Serv. Manag. **19**, 1 (2022). https://doi.org/10.1109/TNSM.2022.3173913
5. Haghi Kashani, M., Mahdipour, E:. Load balancing algorithms in fog computing, 1–18 (2022). https://doi.org/10.1109/TSC.2022.3174475
6. Singh, P., et al.: A fog-cluster based load-balancing technique. Sustainability **14**, 1–14 (2022). https://doi.org/10.3390/su14137961
7. Mahmoudi, Z., Darbanian, E., Nickray, M.: Optimal energy consumption and cost performance solution with delay constraints on fog computing. Jordanian J. Comput. Inf. Technol., 1 (2023). https://doi.org/10.5455/jjcit.71-1667637331
8. Yao, J., Ansari, N.: Fog resource provisioning in reliability-aware IoT networks. IEEE Internet Things J., 1 (2019). https://doi.org/10.1109/JIOT.2019.2922585
9. Klimenko, A.: The basic elements of devices resource consumption decreasing metodology for distributed systems on the basis of fog- and edge-computing. Proc. Southwest State Univ. **26**, 151–167 (2023). https://doi.org/10.21869/2223-15602022-26-3-151-167
10. Melnik, E., Korovin, I., Klimenko, A.: Improving dependability of reconfigurable robotic control system. In: Ronzhin, A., Rigoll, G., Meshcheryakov, R. (eds.) ICR 2017. LNCS (LNAI), vol. 10459, pp. 144–152. Springer, Cham (2017). https://doi.org/10.1007/978-3-319-66471-2_16
11. Korovin, I., Melnik, E., Klimenko, A.: The fog-computing based reliability enhancement in the robot swarm. In: Ronzhin, A., Rigoll, G., Meshcheryakov, R. (eds.) ICR 2019. LNCS (LNAI), vol. 11659, pp. 161–169. Springer, Cham (2019). https://doi.org/10.1007/978-3-030-26118-4_16
12. Klimenko, A.B., Melnik, E.V.: A method of improving the reliability of the nodes containing ledger replicas. In: Silhavy, R., Silhavy, P., Prokopova, Z. (eds.) CoMeSySo 2021. LNNS, vol. 232, pp. 584–592. Springer, Cham (2021). https://doi.org/10.1007/978-3-030-90318-3_47
13. Klimenko, A., Kalyaev, I.: A technique to provide an efficient system recovery in the fog- and edge-environments of robotic systems. In: Ronzhin, A., Rigoll, G., Meshcheryakov, R. (eds.) ICR 2021. LNCS (LNAI), vol. 12998, pp. 100–112. Springer, Cham (2021). https://doi.org/10.1007/978-3-030-87725-5_9
14. Salem, A., Algaphari, G.: Resource allocation in fog computing: a systematic review. J. Sci. Technol. **27**, 9–31. https://doi.org/10.20428/jst.v27i2.2052
15. Gong, C., He, W., Wang, T., Gani, A., Qi, H.: Dynamic resource allocation scheme for mobile edge computing. J. Supercomput. (2023). https://doi.org/10.1007/s11227-023-05323-y
16. Liu, Z., Lan, Q., Huang, K.: Resource allocation for multiuser edge inference with batching and early exiting. IEEE J. Sel. Areas Commun., 1 (2023). https://doi.org/10.1109/JSAC.2023.3242724

17. Murhekar, A., Arbour, D., Mai, T., Rao, A.: Dynamic vector bin packing for online resource allocation in the cloud (2023)
18. Klimenko, A.: Model and method of resource-saving tasks distribution for the fog robotics. In: Ronzhin, A., Meshcheryakov, R., Xiantong, Z. (eds.) ICR 2022. LNCS, vol. 13719, pp. 210–222. Springer, Cham (2022). https://doi.org/10.1007/978-3-031-23609-9_19

Scheduling of Workflows with Task Resource Requirements in Cluster Environments

Oleg Sukhoroslov$^{(\boxtimes)}$ (iD)

Institute for Information Transmission Problems of the Russian Academy of Sciences,
Moscow, Russia
`sukhoroslov@iitp.ru`

Abstract. Workflows is a popular model for computational and data processing applications in science and technology. Such applications are structured as a graph of tasks that run in distributed computing environments such as clusters, grids, and clouds. The choice of algorithm used for scheduling the workflow tasks significantly affects the achieved performance. The proposed workflow scheduling algorithms often use simplified resource allocation models and ignore resource fragmentation effects arising in cluster environments with multicore machines. In this paper we investigate the impact of task resource requirements on performance of several well-known scheduling algorithms in such environments. By means of simulated execution of real scientific workflows we demonstrate that the performance of existing algorithms degrades in the presence of task resource requirements due to inability to efficiently "pack" the tasks in available resources. We propose an alternative approach based on portfolio of heuristics that seeks to strike a balance between the efficient task packing and prioritizing the critical tasks. The proposed approach outperforms the existing algorithms for workflows with irregular task requirements.

Keywords: Distributed computing · workflow · cluster · task scheduling · resource allocation

1 Introduction

Many practical applications executed on parallel and distributed computing environments are structured as a set of loosely coupled tasks with data or control dependencies between them. In particular, computational and data processing pipelines in science and technology are often described and executed in the form of workflows [10]. A workflow task corresponds to execution of a standalone package which reads input data, performs data processing and writes output data. The workflow description defines the data flows between the tasks in the form of producer-consumer relationships. Workflows can be modeled as directed acyclic graphs (DAGs) which vertices correspond to the tasks and edges correspond to the data dependencies between the tasks.

© The Author(s), under exclusive license to Springer Nature Switzerland AG 2023
V. Malyshkin (Ed.): PaCT 2023, LNCS 14098, pp. 177–196, 2023.
https://doi.org/10.1007/978-3-031-41673-6_14

The scientific workflows frequently have high computational demands and therefore are commonly executed in distributed computing environments such as clusters, grids or clouds. The execution of workflows in such environments is automated by means of runtime systems which perform resource management, task scheduling and data movement [11]. An important challenge faced by runtime systems is the task scheduling, i.e. when and on which resource to execute each of the tasks. The scheduling decisions must be made so as to meet the user objectives and constraints, which can encompass notions of performance, monetary cost, energy consumption, reliability, etc. The scheduling decisions are made using a scheduling algorithm which significantly affects the achieved performance in terms of the mentioned objectives. In this work we consider the execution of workflows on clusters of multicore machines and the most popular objective for such environments – minimization of the workflow execution time.

An important practical aspect of workflow execution is the presence of task resource requirements and associated capacity constraints. Indeed, each task requires for its execution a specific number of CPU cores, amount of RAM, disk space and so on. On the other hand, each machine of the target execution environment provides some limited amount of CPU cores, RAM and disk space for the execution of the workflow tasks. The assignment of tasks to machines should be made such as to not exceed the available resource capacity of each machine. Unfortunately, such constraints are often ignored in the literature. Many works on DAG and workflow scheduling use a simple model where each machine is capable of executing any task but only one at each time moment. However, such model is not adequate for modern cluster environments consisting of multicore machines that can have different capacities and can run many tasks simultaneously [43]. In such environments it is necessary to use a more complex resource allocation model that take into account the resource requirements of individual workflow tasks, which can vary from task to task. This aspect is poorly studied in the context of workflow scheduling. While a few works do take into account task resource requirements, they use them only for checking the feasibility of running a task on a machine and do not try to optimize resource allocation.

While the existing workflow scheduling algorithms can be easily modified to respect the additional resource requirements and constraints, it is not clear whether their performance will be maintained in a setting of a cluster of multicore machines. And if not, how the performance can be improved? In this paper we aim to answer these questions by means of simulating the execution of real-world scientific workflows on a diverse set of cluster configurations using several well-known scheduling algorithms. We model different scenarios by varying the task resource requirements in terms of CPU cores and memory. We demonstrate that the performance of critical path-based algorithms degrades in the presence of task resource requirements due to resource fragmentation. We propose an alternative approach based on portfolio of heuristics that seeks to strike a balance between the efficient task packing and prioritizing the critical tasks. The proposed approach outperforms the existing algorithms for workflows with irregular task requirements.

The rest of the paper is organized as follows. Section 2 discusses related work. Section 3 describes the studied workflow scheduling problem, used application and system models, and the evaluated scheduling algorithms. Section 4 presents the evaluation setup, including the used workflow instances, system configurations and algorithm comparison metrics. Section 5 presents and discusses the evaluation results for three scenarios with varying complexity of task packing. Finally, Sect. 6 concludes and outlines the future work.

2   Related Work

The design of algorithms for scheduling workflows or, more generally, task graphs has received an enormous amount of effort. The problem of scheduling a task graph with precedence constraints such as to minimize the makespan has been extensively studied in the context of both homogeneous and heterogeneous multiprocessor systems [22, 36, 41]. This problem is NP-complete in the general case [15], and polynomial-time solutions are known only for a few restricted cases. Therefore researchers have resorted to designing heuristics which can find good solutions within a reasonable amount of time. These heuristics employ various techniques such as list scheduling, task clustering, duplication and guided random search. List scheduling heuristics, based on prioritizing the tasks along the critical path, such as the well-known Heterogeneous Earliest Finish Time (HEFT) algorithm [41], generally provide better results at a lower scheduling time than the other approaches.

The majority of classic DAG scheduling algorithms including HEFT [3, 6, 35, 41] assume that the target system consists of a set of processors capable of executing *any* task but only *one* at each time moment, and that each task is executed by a single processor. The model where processors have different capabilities, i.e. can execute only a subset of the tasks, has been studied in [33]. The basic model has also been refined to account for data-parallel tasks that can be executed on arbitrary numbers of processors [28].

The proliferation of distributed computing environments, clouds and workflow management systems [5, 10] has stimulated a new wave of research related to workflow scheduling algorithms [2, 4, 18, 24, 31]. The studied problem models and assumptions are refined based on the practical aspects of workflow execution in modern computing environments. The problem of efficient workflow execution in grid environments has been addressed by taking into account resource heterogeneity, optimizing network data transfers and using application performance models [7, 26]. The monetary costs associated with task execution in grid and cloud computing environments have been taken into account by extensively studying the related problem statements, including the deadline-constrained [1, 45], budged-constrained [32] and multi-objective [37, 40] formulations. In contrast to previously considered target environments with static configuration, the use of clouds allows to dynamically assemble and manage the task execution environment from the allocated virtual machines (VMs) [27]. More recent works consider the problem of cost-efficient scheduling of multiple workflows in clouds while meeting the diverse QoS requirements [20,25,30,46]. The rising importance of energy efficiency has led to development of workflow scheduling algorithms that explore trade-offs between performance and energy efficiency [13]. The impact of network contention for data-intensive workflows along with the used data placement and transfer strategies are studied in [38,39].

Despite the increased complexity and diversity of the problem settings, most of the proposed workflow scheduling algorithms are based on ideas and principles from the classic DAG scheduling literature and inherit the simple model of resource allocation. As a rule, task resource requirements are ignored and it is naively assumed that any task can be executed on any of the available machines. The simultaneous execution of multiple tasks per machine is also not considered. For example, in a recent work [30] it is assumed that all used VM types have sufficient memory to execute any of the tasks and are only capable of processing one task at a time. A few works that take into account resource requirements use them only for checking the feasibility of running a task on a machine [26,34]. Many works targeting cloud environments avoid resource allocation issues by executing different types of tasks on different VM types with capacities selected based on task requirements [14,30,44]. A common argument against running multiple tasks simultaneously on the same VM is that resource contention can impact the task execution times [44]. However, since VMs actually run on and share resources of physical machines (PMs), this just pushes the resource allocation and isolation issues to the cloud provider which has to map VMs to PMs [14,19].

Despite the rise of cloud computing, clusters of multicore machines is still a popular target environment for workflow execution. In such environments it is a common practice to simultaneously run multiple tasks per machine in order to achieve high resource utilization [43]. Task resource requirements are taken into account during the task scheduling in order to avoid resource shortage and contention. Packing tasks with diverse requirements into machines can lead to such effects as resource fragmentation (the "holes" of unused resources are too small to be useful) and stranding (inability to use idle resources of one type because there are no spare resource of another type) [42]. A common approach to avoid such problems is to employ heuristics for the multidimensional bin packing. This approach has been applied both in the context of task to machine mapping [16] and virtual machine placement [29]. In [17] authors consider scheduling of data-parallel jobs represented as DAGs on clusters. The proposed Graphene scheduler substantially improves the job completion times by leveraging both awareness of task dependencies and efficient task packing. The authors demonstrate that ignoring any of these aspects, i.e. using a critical path-based scheduler or a packing-based scheduler, can lead to poor results for DAGs with heterogeneous resource requirements.

To the best of our knowledge, there are only two recent works that consider scheduling of workflow applications under a similar resource allocation model.

In [47] authors study workflow scheduling in clouds using a model which allows multiple tasks to run concurrently on a VM according to their multi-resource demands. The authors propose a list-scheduling framework for this model and a deadline-constrained workflow scheduling algorithm based on this framework to optimize the cost of workflow execution. However, this work targets a dynamically provisioned cloud-based environment and considers deadline-constrained cost optimization, while our work targets static cluster environment and considers makespan optimization.

In [21] authors consider workflow scheduling on multi-resource clusters with the goal of minimizing the average makespan. They propose GoDAG, an approach that directly learns the scheduling policy from experience through deep reinforcement learning. However, the authors compare GoDAG only with simple Shortest Task First heuristic, classic Dynamic Critical Path algorithm for identical processors [23] and task packing heuristic from [16]. Also, the workflows used in evaluation consist of only 100 tasks and, while using structures from three classic workflow applications, contain randomly generated resource demands and execution times. In contrast, our work includes comparison with several state-of-the-art algorithms, uses a more diverse set of real workflow instances of different sizes and considers multiple scenarios for generating task requirements with increasing complexity. Unfortunately, it is hard to compare our approach with GoDAG because its implementation is not published.

3 Workflow Scheduling

3.1 Problem Statement

Workflow is modeled as a directed acyclic graph (DAG) $G = (T, D)$, where T is the set of tasks and D is the set of dependencies between the tasks. If $(i, j) \in D$ then task t_j depends on some data produced by task t_i, i.e. t_i is a parent task for t_j. A task without parents is called an entry task and a task without children is called an exit task. The edge weight d_{ij} is equal to the amount of data transferred from task t_i to task t_j. A task can be executed on some machine only after all its parent tasks are completed and their outputs are transferred to the machine. It is assumed that tasks are unaware of each other and do not communicate during their execution, i.e., each data transfer occurs only between the executions of the corresponding parent and child tasks.

The tasks are executed in a system represented by a set of machines M connected via a network. Each machine m_k has C_k CPU cores with identical performance and R_k amount of memory. In a homogeneous system all machines have the same resource capacities and performance, while in a heterogeneous system machines' characteristics vary. Each task t_i requires for its execution c_i CPU cores and r_i amount of memory. Each task must be executed entirely on a single machine. The task execution is non-preemptive and can be overlapped with data transfers between the machines. The total amounts of resources (cores and memory) consumed by the tasks executed by the machine at each time moment must not exceed the machine's capacities (C_k and R_k). The execution time $ET(t_i, m_k)$ of each task t_i on each machine m_k can be given explicitly or as follows. Each task has a weight w_i equal to the required amount of computations and each machine has a CPU core performance p_k. Then the task execution time is derived as $ET(t_i, m_k) = w_i/c_i p_k$. We use this approach in the paper by using the number of floating point operations as a task weight and the floating point operations per second as a CPU core performance. The performance degradation due to resource contention between the tasks executing in parallel on the same machine is not modeled.

Before a task t can begin its execution on machine m, its input data (produced by the parents of t) must be transferred to m. When the task is completed its output data is stored on m. When the parent and child tasks are executed on the same machine, the parent's output is available immediately. Otherwise the data must be transferred via the network from the machine that executed the parent task. The time needed to transfer the data between tasks t_i and t_j executed on different machines is computed as $DTT(t_i, t_j) = L + d_{ij}/B$, where L is the network latency and B is the network bandwidth. It is assumed that each data transfer receives the full bandwidth, i.e. there is no contention.

Given the described application and system models, the considered work-flow scheduling problem is to find the assignment of tasks to machines which minimizes the workflow execution time or *makespan*.

3.2   Baseline Algorithms

As a baseline in this work we use several well-known static DAG scheduling algorithms. These algorithms have been applied to workflow scheduling and are also similar to other proposed workflow scheduling algorithms. First, these algorithms are designed for a problem setting with a simple resource allocation model ignoring task resource requirements. Second, all these algorithms are aimed at prioritizing the tasks along the critical path and follow the common iterative list-scheduling technique consisting of two steps. During the task selection step the algorithm selects the next task for scheduling by using some ranking function or criterion. During the resource selection step the algorithm assigns the previously selected task to resource also chosen by using some criterion. These steps are repeated until no further tasks can be scheduled.

Dynamic-level Scheduling (DLS): An algorithm that uses the following *dynamic level* metric computed for each task-machine pair [35]:

$$DL(t_i, m_j) = SL(t_i) + \Delta(t_i, m_j) - EST(t_i, m_j) \ ,$$

where $SL(t_i)$ is the *static level* of task t_i defined as the largest sum of median execution times of tasks along any directed path from t_i to exit task, $\Delta(t_i, n_j)$ is the difference between the median of execution times of task t_i on all machines and $w_{i,j}$, $EST(t_i, m_j)$ is the earliest start time of task t_i on machine m_j. On each iteration the algorithm considers all ready tasks and selects a task-machine pair with the maximum DL value, updates the list of ready tasks and recomputes the DL values.

Heterogeneous Earliest Finish Time (HEFT): A low-complexity heuristic [41] which has demonstrated high efficiency for DAG scheduling and workflow scheduling in particular [39]. The tasks are scheduled in descending order of their rank computed as

$$rank(t_i) = \overline{w_i} + \max_{t_j \in children(t_i)} \left(\overline{DTT(t_i, t_j)} + rank(t_j) \right) ,$$

where $\overline{w_i}$ is the average execution time of task t_i across all machines and $\overline{c_{i,j}}$ is the average data transfer time between tasks t_i and t_j across all pairs of machines. Each task is scheduled to machine with the earliest task finish time. The rank function defines a valid topological order, therefore, similar to DLS, a task is scheduled after its parents and the required earliest start time estimates can be computed.

Lookahead (LA): A static algorithm that can be considered as an extension of HEFT [6]. It uses the same ranking function for task selection, but resource selection is based on scheduling the subsequent tasks using HEFT and selecting machine which minimizes the maximum task finish time. Hence a task completion can be delayed if this reduces the overall makespan, making the algorithm less greedy. The examined subsequent tasks can include only immediate children of a task, recursive children up to some depth, or all remaining tasks. In this work we use the latter variant which has the best performance, but also has the highest computational complexity.

Predict Earliest Finish Time (PEFT): A static algorithm that attempts to achieve the benefits of Lookahead while keeping the computational complexity low [3]. To do this it precomputes the values of Optimistic Cost Table (OCT) for each task-machine pair as follows:

$$OCT(t_i, m_k) = \max_{t_j \in children(t_i)} \min_{m_n \in M} (OCT(t_j, m_n) + ET(t_j, m_n) + DTT(t_i, t_j, m_k, m_n)) .$$

The idea behind this criterion is to estimate the remaining execution time disregarding the machine availability. The tasks are scheduled in decreasing order of the mean OCT value across all machines. A task is assigned to a node that minimizes the sum of task finish time and OCT.

3.3 Accounting for Task Resource Requirements

The described baseline algorithms can be easily adapted to account for task resource requirements and support execution of multiple tasks per machine. Indeed, only the procedure used in machine selection step to find the earliest start time for execution of a task on a given machine should be modified to respect the resource capacity constraints, while the main algorithm logic is not changed.

However, such algorithms can make bad decisions in a setting with varying resource requirements and machine capacities as illustrated on the Fig. 1. The workflow in this example contains four tasks requiring a single CPU core but with different amounts of memory. The system includes two machines with different processing speeds and resource capacities. The schedule produced by HEFT (also by DLS) is depicted on the upper right corner. Task B is scheduled before C and

D, and is assigned to the fastest machine, leaving no possibility for simultaneous execution of D due to the memory fragmentation. The optimal schedule depicted on the bottom right corner places B on the slower machine thereby allowing the simultaneous execution of all three tasks.

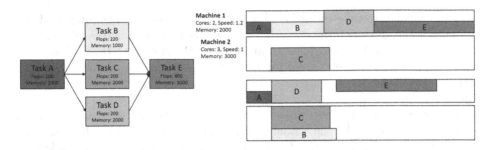

Fig. 1. Suboptimal scheduling of tasks with resource requirements by HEFT.

This example illustrates the motivation behind this work. HEFT and other baseline algorithms are aimed at optimizing the execution of tasks along the critical path. But in the presence of shared resources such as CPU cores and memory an uncareful task assignment can lead to resource fragmentation that can block the execution of subsequent tasks. This results in reduced system utilization and increased workflow execution time. In this particular example, Lookahead and PEFT manage to produce the optimal schedule. However, similar "bad" examples can be provided for these algorithms as well. Also note that the presented example showed the case of only a single resource (memory) fragmentation. When the tasks have different CPU cores requirements the situation becomes more complex since both resources can be fragmented.

3.4   Heuristics Portfolio

To improve the workflow scheduling in a setting with varying resource requirements and machine capacities we propose to leverage the multidimensional bin packing heuristics. Such heuristics has been previously applied to improve resource utilization and avoid fragmentation in the context of cluster scheduling [16] and virtual machine placement [29]. Inspired by the classic bin packing heuristics such as BestFit, such algorithms assign a task to machine that maximizes some packing score. For example, in [16] the best packing efficiency was obtained by using as such score the dot product between the vector of task resource requirements and the vector of available resources on the machine.

Instead of relying on a single heuristic we propose to use a set of heuristics that seeks to strike a balance between the efficient task packing and prioritizing the critical tasks. These heuristics are based on two list scheduling variants and use different task and machine selection criteria. The schedule is build dynamically in runtime by considering only ready tasks and idle resources, though a similar approach can also be applied as a static algorithm.

The first list scheduling variant is the same as in the baseline algorithms – the task selection step is followed by the machine selection step. The task selection step considers only ready tasks (i.e. tasks which parents are already completed) and uses one of the following criteria (ties are broken by preferring the task with maximum flops):

1. Pick the task with the largest amount of computations (flops),
2. Pick the task with the largest required CPU cores,
3. Pick the task with the largest required memory,
4. Pick the task with the largest sum of normalized CPU cores and memory,
5. Pick the task with the largest product of normalized CPU cores and memory,
6. Pick the task with the largest product of normalized CPU cores and flops,
7. Pick the task with the largest product of normalized memory and flops,
8. Pick the task with the largest product of flops and the sum of normalized CPU cores and memory,
9. Pick the task with the largest rank as computed in HEFT,
10. Pick the task with the largest amount of input and output data,
11. Pick the task with the largest number of children tasks.

The machine selection step considers only machines with enough idle resources and uses one of the following criteria (ties are broken by preferring the machine with the highest speed):

1. Pick the machine with the highest speed,
2. Pick the machine with the most available CPU cores,
3. Pick the machine with the most available memory,
4. Pick the machine with the least available CPU cores,
5. Pick the machine with the least available memory,
6. Pick the machine with the largest dot product between task requirements and available resources (normalized to range from 0 to 1),
7. Pick the machine with the highest sum of dot product (see above) and normalized machine speed,
8. Pick the machine that stores the largest amount of task input data.

The second list scheduling variant reverses the steps – the machine selection step is followed by the task selection step. The idea is to pick the most suitable task to pack into the given machine. The machine selection step currently uses a single criterion – pick the machine with the fastest speed among the machines with available resources (ties are broken by preferring the machine with the most available CPU cores). The task selection step considers only ready tasks and uses one of the following criteria (ties are broken by preferring the task with highest flops):

1. Pick the task with the largest task rank as computed in HEFT,
2. Pick the task with the largest dot product between task requirements and available resources,

3. Pick the task with the largest weighted sum of normalized task rank and dot product (as above): $score(t_i, m_k) = \alpha rank(t_i) + (1 - \alpha)dot\_product(t_i, m_k)$, where α regulates the priority of critical path scheduling in comparison to efficient task packing (values 0.25, 0.5 and 0.75 were used in experiments),
4. Pick the task with the largest product of task rank and dot product,
5. Pick the task with the largest dot product among the top N tasks by rank (N=10 was used in experiments).

This results in 95 heuristics in total implementing different list scheduling variants and criteria combinations. We combine these heuristics in a single portfolio algorithm which takes the best schedule found by the heuristics.

4   Evaluation Setup

4.1   Workflow Instances

In this work we use 150 real-world scientific workflow instances corresponding to 9 applications from different domains such as bioinformatics, seismology, astronomy, etc. These instances, derived from the logs of actual workflow executions, are provided by the WfCommons project [9]. Each workflow instance is provided as a JSON file conforming to the WfCommons JSON Schema. It includes the information about the execution environment (machines, their speed and number of cores) and the executed workflow tasks (task name, task runtime, input and output files with their sizes, machine executed the task). This information allows to recover the DAG structure and the weights of its vertices and edges according to the previously described model.

Table 1. Characteristics of workflow instances

Workflow	#	Tasks	Depth	Width	Parallelism	CCR	Max work/data
1000Genome	22	52–902	3	28–572	13–152	47–487	2.14/75.62
BLAST	15	43–303	3	40–300	34–269	0.02–6.5e5	42.87/5119
BWA	15	104–1004	3	100–1000	4–26	32–105	3.78/57
Cycles	16	67–1091	4	32–540	5–62	2265–7741	0.9/7.77
Epigenomics	26	41–1695	9	9–420	5–97	684–2373	1.17/12.7
Montage	11	58–1312	8	18–936	10–79	234–1055	3.08/17.42
Seismology	10	101–1001	2	100–1000	27–116	36694–78748	0.02/0.02
SoyKB	10	96–676	11	50–500	2.5–7.8	62–174	6.03/2.88
SRA Search	25	22–104	3–4	11–51	5–24	775–3092	5.64/78.87

Table 1 summarizes the characteristics of the used workflow instances. The small DAG depths and large widths indicate that these workflows have high potential for parallel execution. The parallelism degree is evaluated based on the ratio of the total amount of computations (task sizes) and the critical path length (omitting the data transfers). The computation-to-communication ratio (CCR), defined as the ratio of the total amount of computations (in Gflops) and the total size of data transfers (in GBytes), allows to estimate the impact of data transfers and identify data intensive workflows. Finally we report the maximum values of work (hours of execution on 10 Gflop/s machine) and total files size in GB for each workflow type.

Unfortunately, the CPU cores and memory used by tasks are provided only for BLAST and BWA workflows. Therefore in this work we use randomly generated task requirements. The three versions of each workflow are created which correspond to different scenarios with increasing task packing difficulty:

- *Simple*: each task requires a single CPU core and has no memory requirements.
- *Regular*: all tasks of the same type (inferred by the task name prefix) have the same randomly generated CPU and memory requirements.
- *Irregular*: each task has individual randomly generated CPU and memory requirements.

The CPU cores requirements are generated in the range $[1, min_k(C_k)]$, while the memory requirements are generated in the range $[2, min_k(R_k)]$ GBytes. Only integer values were used for both requirements.

4.2   System Configurations

The scientific workflows are frequently executed on clusters consisting of multicore machines connected with Ethernet network. To provide a diverse set of scenarios for experiments we used 8 cluster configurations listed in Table 2. These configurations differ by their size, heterogeneity and performance. Each machine has from 4 to 24 cores with performance ranging from 2 to 6 Gflop/s (this is not a theoretical peak performance but the actual performance achieved for executed tasks). The machines are connected with 100 GbE network.

Table 2. System configurations

System	Machines	Cores per Machine	Memory per Machine
cluster-hom-4-32	4	8	16
cluster-hom-4-64	4	16	32
cluster-hom-8-64	8	8	16
cluster-hom-8-128	8	16	32
cluster-het-4-32	4	4,8,16	8,16,32
cluster-het-4-64	4	8,12,20,24	16,24,40,48
cluster-het-8-64	8	4,8,16	8,16,32
cluster-het-8-128	8	8,16,24	16,32,48

4.3   Performance Metrics

The following metrics are used for comparison of scheduling algorithms.

The main performance measure of a scheduling algorithm s on a particular problem instance $p = (workflow, system)$ is $makespan_{s,p}$ - the workflow execution time under the schedule produced by the algorithm. However, since a large set of workflow instances with diverse properties (e.g. critical path length) is used, it is necessary to normalize the makespan to meaningfully compare and aggregate the results across all instances. For this purpose we use the Degradation from Best (DFB) metric [8] which is the relative difference between the algorithm's makespan and that achieved by the best algorithm for this problem instance:

$$DFB(s,p) = \frac{makespan_{s,p} - \min\{makespan_{s,p} : s \in S\}}{\min\{makespan_{s,p} : s \in S\}}.$$

An alternative way to compare the results of multiple algorithms across a diverse set of problem instances relies on performance profiles [12]. Originally introduced to assess the performance of optimization software, performance profiles provide a general method to aggregate and visualize the benchmark results. Algorithm's performance profile is the cumulative distribution function for the performance ratio

$$\rho_s(\tau) = \frac{1}{n_p}\text{size}\{p \in P : r_{p,s} \le \tau\},$$

where $n_p = |P|$ is the number of problem instances, $r_{p,s}$ is the performance ratio of algorithm s on problem instance p:

$$r_{p,s} = \frac{makespan_{s,p}}{\min\{makespan_{s,p} : s \in S\}}.$$

We also measure and report the running time of an algorithm, i.e. its execution time for computing the schedule for a given workflow. This metric corresponds to the cost of using the algorithm.

4.4   Simulation Framework

The evaluation of scheduling algorithms on the used workflow instances is performed by means of simulation. For each $(workflow, system, algorithm)$ triple we simulate the execution of the workflow in the system using the schedule produced by the algorithm. As a result, we obtain the achieved makespan, i.e. the duration of workflow execution. It is assumed that the task execution and data transfer times used by the algorithms are accurate, i.e. they are not changed during the simulation. Therefore the obtained makespan should be equal to the makespan expected by the algorithm.

For simulation purposes we use DSLab DAG[1], a library for studying the DAG scheduling algorithms. It allows to describe a DAG and simulate its execution

---

[1] https://github.com/osukhoroslov/dslab/tree/main/crates/dslab-dag.

in a given distributed system using the specified scheduling algorithm, including the user-defined one. The library includes the implementations of the considered scheduling algorithms.

5   Evaluation Results

Using DSLab DAG we have performed 118800 simulations (99 algorithms × 150 workflow instances × 8 systems) for each considered scenario.

5.1   Simple Scenario

In this scenario we consider workflow instances where each task requires a single CPU core and has no memory requirements. This is the most simple case in terms of task packing which sets the baseline for further scenarios.

The aggregated results for each algorithm are presented in Table 3. We report the number of times the algorithm produced the best makespan and the average values of DFB and running time (80th percentile values are also provided in brackets). The corresponding performance profiles are presented on Fig. 2 (left). As it can be seen, all baseline algorithms except PEFT clearly outperform Portfolio. Lookahead performs the best but with significantly higher running time, while HEFT and DLS have a similar performance in terms of makespan.

Among the Portfolio algorithms the best results are achieved by the following heuristics that outperform PEFT:

- Pick the task with the largest HEFT rank, schedule the task to machine with the highest sum of normalized dot product and machine speed,
- Pick the machine with the fastest speed, schedule to it the task chosen using one of criteria combining dot product and HEFT rank,
- Pick the task with the largest HEFT rank, schedule the task to machine with the fastest speed.

The result of the best heuristic is also presented in the Table 3. From there it can be seen that the combination of multiple heuristics into Portfolio indeed allows to significantly improve the performance since different heuristics contribute best results for different cases.

5.2   Regular Scenario

In this scenario we consider workflow instances where all tasks of the same type (inferred by the task name) have the same randomly generated CPU and memory requirements. This is a reasonable assumption if the tasks of the same type comprise a single workflow stage and process the similarly sized inputs. Since the used workflow instances have small depths and large widths (see Table 1), this scenario produces many identical tasks with only a small variation of their "sizes".

Table 3. Results for simple scenario.

Algorithm	Best	DFB, %	Running time
Lookahead	1104	0.04 (0)	365 (330)
HEFT	541	0.61 (1.15)	0.07 (0.10)
DLS	555	0.80 (1.41)	8.37 (6.85)
PEFT	273	8.68 (15.58)	0.08 (0.11)
Portfolio	46	3.77 (4.87)	10.64 (14.11)
Best from Portfolio	11	6.76 (8.38)	3.84 (1.73)

The aggregated results for each algorithm are presented in Table 4. The performance profiles are presented on Fig. 2 (right). It can be seen that the relative performance of the algorithms has noticeably changed. While Lookahead and HEFT still have the best performance, the gap between them and Portfolio has significantly decreased. At the same time DLS performance is degraded and became close to PEFT. This demonstrates that the task resource requirements indeed impact the performance of baseline algorithms, while Portfolio takes advantage from using the packing heuristics. It outperforms Lookahead by more than 10% in 1.8% of cases, with the largest relative gap of 34%.

The best results within Portfolio are indeed achieved by heuristics using the second list scheduling variant and combining the dot product with HEFT ranks. The best heuristic contributes only about 17% of the best Portfolio results which again confirms the advantage of combining multiple heuristics.

Table 4. Results for regular scenario.

Algorithm	Best	DFB, %	Running time
Lookahead	962	0.70 (0)	801 (706)
HEFT	81	1.66 (2.50)	0.09 (0.13)
DLS	58	5.88 (7.74)	8.83 (9.95)
PEFT	30	6.54 (10.53)	0.09 (0.13)
Portfolio	187	2.76 (2.87)	14.42 (17.91)
Best from Portfolio	31	9.27 (11.27)	0.60 (0.18)

5.3   Irregular Scenario

In this scenario we consider workflow instances where each task has individual randomly generated CPU and memory requirements. This is the most challenging scenario in terms of the task packing since the produced task "sizes" are highly irregular. This may correspond to a situation when the input data sizes vary between the tasks of the single type.

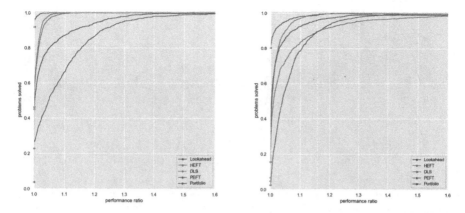

Fig. 2. Performance profiles for simple (left) and regular (right) scenarios.

The aggregated results for all algorithms are presented in Table 5. The performance profiles are presented on Fig. 3 (left). As expected, the increase of task requirements heterogeneity leads to further degradation of the baseline algorithms due to inefficient task packing and resource fragmentation. Portfolio is the clear leader in this scenario in terms of the average and 80th percentile performance. It outperforms Lookahead by more than 10% in 14% of cases, with the largest relative gap of 46%. On the other hand, in 8.7% of cases Portfolio is outperformed by the baseline algorithms by more than 10%, with the largest relative gap of 85%. These observations are aligned with the relative positions of the performance profile curves. Note also the prohibitively high running times of Lookahead that are further increased in this scenario.

Table 5. Results for irregular scenario.

Algorithm	Best	DFB, %	Running time
Lookahead	280	4.88 (8.37)	1065 (908)
HEFT	44	8.42 (12.89)	0.09 (0.14)
DLS	37	9.97 (14.00)	10.06 (11.33)
PEFT	29	12.11 (17.70)	0.09 (0.15)
Portfolio	858	3.15 (1.18)	13.80 (16.06)
Best from Portfolio	47	21.29 (37.39)	0.77 (0.18)

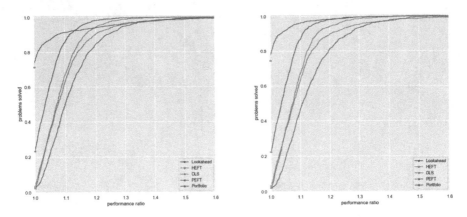

Fig. 3. Performance profiles for irregular scenario (left - all applications, right - without 1000Genome and BLAST).

The best results within Portfolio are again achieved by heuristics using the second list scheduling variant and combining the dot product with HEFT ranks. However now the best heuristic contributes only about 5% of the best Portfolio results, so it strengthens the argument for using the portfolio approach.

The performance profiles for irregular scenario per each workflow application are presented on Fig. 4. For the most of applications, except 1000Genome and BLAST, there is consistent advantage of Portfolio, while the gap varies from application to application. The performance profiles after excluding these two applications are presented on Fig. 3 (right). The least advantage of Portfolio is observed for 1000Genome, where it is outperformed by the baseline algorithms by more than 10% in 20% of cases. As for BLAST, Portfolio achieves the best average DFB but in some cases its makespan is significantly larger than Lookahead's one up to 85%. This clearly shows a room for improvement of the proposed approach which will be investigated in the future work.

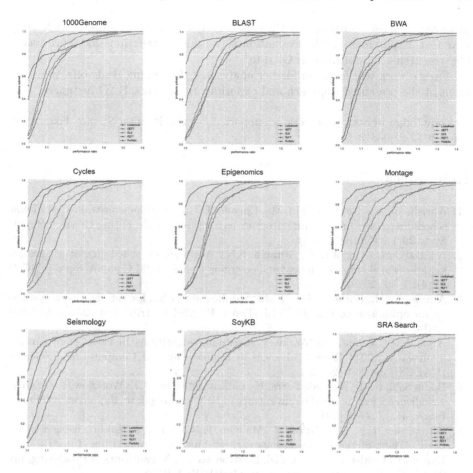

Fig. 4. Performance profiles for irregular scenario per application.

6   Conclusion and Future Work

In this paper we have investigated the impact of task resource requirements and related constraints on the performance of workflow scheduling algorithms in cluster environments with multicore machines capable of executing multiple tasks simultaneously. By means of simulated execution of real scientific workflows on a diverse set of cluster configurations we have demonstrated that the performance of existing algorithms degrades in the setting with varying resource requirements and machine capacities. This can be explained by the fact that such algorithms, aimed at prioritizing the tasks along the critical path, can make poor allocation decisions leading to resource fragmentation and reducing the opportunities for the execution of subsequent tasks. To improve the scheduling performance in such settings we have proposed the portfolio of heuristics that seeks to strike a balance between the efficient task packing and prioritizing the critical tasks. The proposed approach outperforms the existing algorithms for workflows with irregular task requirements. The evaluation setup and instructions for reproducing this research are published on GitHub[2].

Future work will focus on further analysis of the arising trade-offs, improvement of the presented approach and experiments on other DAG instances.

Acknowledgements. This work is supported by the Russian Science Foundation (project 22-21-00812).

References

1. Abrishami, S., Naghibzadeh, M., Epema, D.H.: Deadline-constrained workflow scheduling algorithms for infrastructure as a service clouds. Future Gener. Comput. Syst. **29**(1), 158–169 (2013)
2. Adhikari, M., Amgoth, T., Srirama, S.N.: A survey on scheduling strategies for workflows in cloud environment and emerging trends. ACM Comput. Surv. (CSUR) **52**(4), 1–36 (2019)
3. Arabnejad, H., Barbosa, J.G.: List scheduling algorithm for heterogeneous systems by an optimistic cost table. IEEE Trans. Parallel Distrib. Syst. **25**(3), 682–694 (2014)
4. Arya, L.K., Verma, A.: Workflow scheduling algorithms in cloud environment- a survey. In: 2014 Recent Advances in Engineering and Computational Sciences (RAECS), pp. 1–4 (2014)
5. Badia Sala, R.M., Ayguadé Parra, E., Labarta Mancho, J.J.: Workflows for science: a challenge when facing the convergence of HPC and big data. Supercomput. Front. Innov. **4**(1), 27–47 (2017)
6. Bittencourt, L.F., Sakellariou, R., Madeira, E.R.M.: Dag scheduling using a lookahead variant of the heterogeneous earliest finish time algorithm. In: 2010 18th Euromicro Conference on Parallel, Distributed and Network-based Processing, pp. 27–34 (Feb 2010). https://doi.org/10.1109/PDP.2010.56
7. Blythe, J., et al.: Task scheduling strategies for workflow-based applications in grids. In: CCGrid 2005 IEEE International Symposium on Cluster Computing and the Grid, 2005. vol. 2, pp. 759–767. IEEE (2005)
8. Casanova, H., Wong, Y.C., Pottier, L., da Silva, R.F.: On the feasibility of simulation-driven portfolio scheduling for cyberinfrastructure runtime systems. In: Job Scheduling Strategies for Parallel Processing (2022)
9. Coleman, T., Casanova, H., Pottier, L., Kaushik, M., Deelman, E., da Silva, R.F.: WfCommons: a framework for enabling scientific workflow research and development. Future Gener. Comput. Syst. **128**, 16–27 (2022)
10. Deelman, E., Gannon, D., Shields, M., Taylor, I.: Workflows and e-Science: an overview of workflow system features and capabilities. Future Gener. Comput. Syst. **25**(5), 528–540 (2009)
11. Deelman, E., et al.: Pegasus, a workflow management system for science automation. Future Gener. Comput. Syst. **46**, 17–35 (2015)
12. Dolan, E.D., Moré, J.J.: Benchmarking optimization software with performance profiles. Math. programm. **91**, 201–213 (2002)

---

[2] https://github.com/osukhoroslov/pact2023-experiments.

13. Durillo, J.J., Nae, V., Prodan, R.: Multi-objective energy-efficient workflow scheduling using list-based heuristics. Future Gener. Comput. Syst. **36**, 221–236 (2014)
14. Esteves, S., Veiga, L.: WaaS: workflow-as-a-service for the cloud with scheduling of continuous and data-intensive workflows. Comput. J. **59**(3), 371–383 (2016)
15. Garey, M.R., Johnson, D.S.: Computers and intractability, vol. 174. freeman San Francisco (1979)
16. Grandl, R., Ananthanarayanan, G., Kandula, S., Rao, S., Akella, A.: Multi-resource packing for cluster schedulers. ACM SIGCOMM Comput. Commun. Rev. **44**(4), 455–466 (2014)
17. Grandl, R., Kandula, S., Rao, S., Akella, A., Kulkarni, J.: Graphene: packing and dependency-aware scheduling for data-parallel clusters. In: Proceedings of the 12th USENIX Conference on Operating Systems Design and Implementation, pp. 81–97. OSDI'16, USENIX Association, USA (2016)
18. Gupta, A., Garg, R.: Workflow scheduling in heterogeneous computing systems: A survey. In: 2017 International Conference on Computing and Communication Technologies for Smart Nation (IC3TSN), pp. 319–326. IEEE (2017)
19. Hadary, O., et al.: Protean: VM allocation service at scale. In: Proceedings of the 14th USENIX Conference on Operating Systems Design and Implementation, pp. 845–861 (2020)
20. Hilman, M.H., Rodriguez, M.A., Buyya, R.: Multiple workflows scheduling in multi-tenant distributed systems: a taxonomy and future directions. ACM Comput. Surv. (CSUR) **53**(1), 1–39 (2020)
21. Hu, Y., de Laat, C., Zhao, Z.: Learning workflow scheduling on multi-resource clusters. In: 2019 IEEE International Conference on Networking, Architecture and Storage (NAS), pp. 1–8. IEEE (2019)
22. Kwok, Y.K., Ahmad, I.: Benchmarking the task graph scheduling algorithms. In: Proceedings of the first merged international parallel processing symposium and symposium on parallel and distributed processing, pp. 531–537. IEEE (1998)
23. Kwok, Y.K., Ahmad, I.: Dynamic critical-path scheduling: an effective technique for allocating task graphs to multiprocessors. IEEE Trans. Parallel Distrib. Syst. **7**(5), 506–521 (1996)
24. Liu, J., Pacitti, E., Valduriez, P., Mattoso, M.: A survey of data-intensive scientific workflow management. J. Grid Comput. **13**, 457–493 (2015)
25. Malawski, M., Juve, G., Deelman, E., Nabrzyski, J.: Algorithms for cost-and deadline-constrained provisioning for scientific workflow ensembles in IaaS clouds. Future Gener. Comput. Syst. **48**, 1–18 (2015)
26. Mandal, A., et al.: Scheduling strategies for mapping application workflows onto the grid. In: HPDC-14. Proceedings. 14th IEEE International Symposium on High Performance Distributed Computing, 2005, pp. 125–134. IEEE (2005)
27. Mao, M., Humphrey, M.: Auto-scaling to minimize cost and meet application deadlines in cloud workflows. In: Proceedings of 2011 International Conference for High Performance Computing, Networking, Storage and Analysis, pp. 1–12 (2011)
28. N'takpé, T., Suter, F., Casanova, H.: A comparison of scheduling approaches for mixed-parallel applications on heterogeneous platforms. In: Sixth International Symposium on Parallel and Distributed Computing (ISPDC'07), pp. 35–35. IEEE (2007)
29. Panigrahy, R., Talwar, K., Uyeda, L., Wieder, U.: Heuristics for vector bin packing. http://research.microsoft.com (2011)

30. Rodriguez, M.A., Buyya, R.: Scheduling dynamic workloads in multi-tenant scientific workflow as a service platforms. Future Gener. Comput. Syst. **79**, 739–750 (2018)
31. Rodriguez, M.A., Buyya, R.: A taxonomy and survey on scheduling algorithms for scientific workflows in IaaS cloud computing environments. Concurrency Comput.: Pract. Experience **29**(8), e4041 (2017)
32. Sakellariou, R., Zhao, H., Tsiakkouri, E., Dikaiakos, M.D.: Scheduling workflows with budget constraints. In: Integrated Research in GRID Computing: CoreGRID Integration Workshop 2005 (Selected Papers) November 28–30, Pisa, Italy, pp. 189–202. Springer (2007). https://doi.org/10.1007/978-0-387-47658-2_14
33. Shi, Z., Dongarra, J.J.: Scheduling workflow applications on processors with different capabilities. Future Gener. Comput. Syst. **22**(6), 665–675 (2006)
34. Shrestha, H., et al.: Scheduling workflows on a cluster of memory managed multi-core machines. In: Arabnia, H.R. (ed.) Proceedings of the International Conference on Parallel and Distributed Processing Techniques and Applications, PDPTA 2009, Las Vegas, Nevada, USA, July 13–17, 2009, vol. 2 , pp. 631–637. CSREA Press (2009)
35. Sih, G.C., Lee, E.A.: A compile-time scheduling heuristic for interconnection-constrained heterogeneous processor architectures. IEEE Trans. Parallel Distrib. Syst. **4**(2), 175–187 (1993)
36. Sinnen, O.: Task scheduling for parallel systems. John Wiley & Sons (2007)
37. Su, S., Li, J., Huang, Q., Huang, X., Shuang, K., Wang, J.: Cost-efficient task scheduling for executing large programs in the cloud. Parallel Comput. **39**(4–5), 177–188 (2013)
38. Sukhoroslov, O.: Toward efficient execution of data-intensive workflows. J. Supercomput. **77**(8), 7989–8012 (2021). https://doi.org/10.1007/s11227-020-03612-4
39. Sukhoroslov, O., Nazarenko, A., Aleksandrov, R.: An experimental study of scheduling algorithms for many-task applications. J. Supercomput. **75**, 7857–7871 (2019)
40. Szabo, C., Kroeger, T.: Evolving multi-objective strategies for task allocation of scientific workflows on public clouds. In: 2012 IEEE Congress on Evolutionary Computation, pp. 1–8. IEEE (2012)
41. Topcuoglu, H., Hariri, S., Wu, M.Y.: Performance-effective and low-complexity task scheduling for heterogeneous computing. IEEE Trans. Parallel Distrib. Syst. **13**(3), 260–274 (2002). https://doi.org/10.1109/71.993206
42. Verma, A., Korupolu, M., Wilkes, J.: Evaluating job packing in warehouse-scale computing. In: 2014 IEEE International Conference on Cluster Computing (CLUSTER), pp. 48–56. IEEE (2014)
43. Verma, A., Pedrosa, L., Korupolu, M., Oppenheimer, D., Tune, E., Wilkes, J.: Large-scale cluster management at Google with Borg. In: Proceedings of the Tenth European Conference on Computer Systems, pp. 1–17 (2015)
44. Wang, J., Korambath, P., Altintas, I., Davis, J., Crawl, D.: Workflow as a service in the cloud: architecture and scheduling algorithms. Procedia Comput. Sci. **29**, 546–556 (2014)
45. Yu, J., Buyya, R., Tham, C.K.: Cost-based scheduling of scientific workflow applications on utility grids. In: First International Conference on e-Science and Grid Computing (e-Science'05), p. 8. IEEE (2005)
46. Zhou, A.C., He, B., Liu, C.: Monetary cost optimizations for hosting workflow-as-a-service in IaaS clouds. IEEE Trans. Cloud Comput. **4**(1), 34–48 (2015)
47. Zhu, Z., Tang, X.: Deadline-constrained workflow scheduling in IaaS clouds with multi-resource packing. Future Gener. Comput. Syst. **101**, 880–893 (2019)

Verifying the Correctness of HPC Performance Monitoring Data

Danil Kashin[1] and Vadim Voevodin[2]([⊠]) [ID]

[1] Faculty of Computational Mathematics and Cybernetics,
Lomonosov Moscow State University, Moscow 119234, Russia
[2] Research Computing Center, Lomonosov Moscow State University,
Moscow 119234, Russia
vadim@parallel.ru

Abstract. Administration and maintenance of modern supercomputers requires monitoring not only the correctness of their work, but also the efficiency of their functioning. For these purposes, monitoring systems are used that constantly run on supercomputer nodes and collect information about how actively and efficiently these nodes are used. The analysis of such information allows you, for example, to investigate the performance of individual jobs or users, study the efficiency of application packages usage, analyze the utilization of service nodes, or compare the job behavior in different partitions. However, for this to be possible, it is necessary to be sure that the performance data collected by the monitoring system is correct. One way to check the correctness of such data is to use a set of external tests. Each test, when executed on a supercomputer, gives the expected value for some performance characteristic like CPU user load, I/O read speed or the frequency of L1 cache misses, which is also being collected by the monitoring system. The matching of the value expected by such test and the value collected by the monitoring system indicates that the general process of handling this performance data is implemented correctly. This paper provides a description of such open-source test suite developed at the Lomonosov Moscow State University.

Keywords: Supercomputing · Performance monitoring · Monitoring data · Performance analysis · Test suite · Correctness

1 Introduction

Modern supercomputers consist of a huge number of various software and hardware components, each of which can start working incorrectly or even fail. Therefore, it is necessary to constantly monitor the current state of the supercomputer and its components in order to be able to quickly identify and eliminate such cases of malfunction. However, it is important to constantly monitor not only the availability and correct operation of the equipment, but also the efficiency of its usage. For these purposes, administrators of supercomputing centers use

V. Malyshkin (Ed.): PaCT 2023, LNCS 14098, pp. 197–208, 2023.
https://doi.org/10.1007/978-3-031-41673-6_15

a performance monitoring system – a type of monitoring system that is capable of or even mainly focused on collecting performance data describing CPU load, GPU memory usage or frequency of memory load operations, etc. Such systems usually have individual agents running on compute and service nodes of the supercomputer, which constantly collect various information about node utilization and then transfer it to the system servers for aggregation and storage. This allows jointly analyzing all the collected performance-related information about the supercomputer using, for example, specific analytical systems like XDMoD [17], OMNI [9] or TASC [18].

At any stage of processing performance monitoring data (collection, transfer, transformation, aggregation, storing), a mistake can be made, which can lead to incorrect output data. And such mistake can occur both within the performance monitoring system itself and outside of it, for example, within the analytical system or during data output (see examples in Sect. 3). At the same time, it is often difficult for supercomputer administrators to understand whether the performance data provided is adequate. For example, according to the collected data, it is shown that during the execution of a certain job, the average L1 cache miss rate was equal to 10 million misses per second. Is this value correct? External verification tests can help answer such questions, and their development was the main goal of this work.

The main contribution of this paper is the proposal of the software test suite[1] that is designed to check the correctness of the output monitoring data used for performance analysis of supercomputer functioning. Each test in this suite is executed on a supercomputer in a usual user mode, using the queuing system. After its completion, it produces the expected value for some performance characteristic on one compute node, which can then be compared with the value collected by the monitoring system itself (and optionally further processed, e.g. by the analytical system). If the values match, then the monitoring process is configured correctly in general; if the values differ notably, this indicates the presence of an error at one of the data processing stages. Current version of this suite include tests for 10 commonly used performance characteristics describing CPU and GPU load, MPI usage, cache miss frequency and amount of free memory. More tests can be easily added using the provided mechanisms, which is planned to be done in the future.

The usage of this solution is of the greatest interest during the initial setup and launch of the system software needed for supercomputer performance analysis – monitoring, analytical and data visualization systems. However, during the operation of the supercomputer, modifications are also sometimes made in the process of working with performance monitoring data, and then such a check should also be performed. Moreover, it is useful, in our opinion, to periodically run such checks to make sure that nothing accidentally has happened to the data handling process.

---

[1] Available at https://github.com/KashinDanil/JDC.

The rest of the paper is organized as follows. Section 2 describes existing works related to the topic of this paper. Section 3 explains why performance data verification if important and relevant. Section 4 describes the proposed test suite itself, while Sect. 5 is devoted to the evaluation of the test suite on two modern supercomputers. Conclusions and future plans are described in Sect. 6.

2   Related Work

The topic discussed in this paper primarily relates to the task of supercomputer performance monitoring. Therefore, we studied different existing systems for monitoring and collecting supercomputer data to find out if they provide mechanisms (test suites or similar) to check the correctness of the data collected. We have considered such well-known general monitoring systems as Zabbix [16], Nagios [8] and Ganglia [14]; performance monitoring systems SuperMon [19], OVIS-2 [10], TACC Stats [11], Dataheap [13] and Distributed Modular Monitoring [20]; and LIKWID performance tool suite [12].

Among these solutions, not a single one was found that somehow solve this task. In our opinion, this may be due to the following reasons. Firstly, developers of monitoring systems are primarily interested in ensuring that their particular system works correctly, and such checks can be largely carried out at the development stage. While we want to create a universal solution that works with different system software and to verify the data not only at the output of the monitoring system, but also after subsequent data processing or after the work of analytical systems. Secondly, monitoring systems are usually focused on collecting data on the availability and operability of supercomputers, and this data, unlike performance data, is often notably easier to verify.

As mentioned earlier, each of the proposed tests within the test suite is designed to verify the correctness of the monitoring data collected for particular performance characteristic like CPU load or frequency of L1 cache misses. In this case, individual tests can be developed independently. And although we did not find ready-made solutions that solve the stated task, we found existing benchmarks and tests, designed for other purposes, that can be used in some of the proposed tests.

The first such solution is HPAS (HPC Performance Anomaly Suite) [7]. This is a software tool designed to generate performance anomalies to check the impact of such anomalies on the behavior of parallel programs and the system as a whole. The tool includes various types of anomalies, such as CPU load, network delays, usage of RAM and I/O, and others. Another solution that we used is the OSU Micro-benchmarks test suite [3], designed to evaluate the performance of various MPI operations (not only MPI in the latest versions) on a computational system. The set contains many different tests that can be used to evaluate data and packet transfer rates as well as latency, when using collective, point-to-point and one-sided MPI operations. In Sect. 4, it will be shown how these solutions were applied.

3   Motivation for Performance Data Verification

In this section, we will explain why it is important to check the correctness of performance monitoring data collected on the supercomputer. We will discuss two related questions – why validation of performance data requires special attention, and also why such data can be incorrect in practice.

Performance data, unlike most data on the correctness and availability of supercomputer components (which is mainly collected in practice using monitoring systems), is very difficult to verify without specific verification tools. Sometimes it is even difficult to determine whether the order of magnitude is correct. For example, it is shown that the average number of performed instructions is 3 billions per second, or the amount of data transferred over the MPI network is 300 MB per second. How to be sure that these values are correct? Even knowing which program is being executed, it is often difficult to answer this question. And without focused checks on the correctness of such data, a noticeable time can pass before errors in this data are discovered. At the same time, the price of such errors is high, since incorrect data can easily lead to completely wrong conclusions about the performance of user applications or the supercomputer in general. Therefore, in our opinion, it is necessary to pay special attention to the verification of performance data.

Now, let us explain why errors in performance monitoring data can occur in practice. Speaking in general, there are several points worth noting. Monitoring systems can be configured in different ways that can sometimes impact the data correctness. Moreover, it is required after data collection to organize its transfer and storage, as well as pre-processing and aggregation in some cases. And these stages are often performed not by the monitoring system, but by external tools like analytical systems or data visualization tools, which increases the likelihood of inaccuracies in the overall process of data handling.

Let us consider several real-life cases that we encountered in our practice, when there was a need for performance data verification. The presence of these cases inspired us to develop the proposed test suite, in order to make their detection and elimination a much easier task in future.

Processors sometimes have errors in their design. For example, in some processors of the Intel Haswell and Ivy Bridge series, correct data collection from different hardware memory-related counters is not guaranteed when using Simultaneous Multithreading, or SMT (for example, see errata HSW29 [6] or CA93 [5]). This includes counters that can be useful in performance analysis, such as those that estimate the number of memory read and write operations or the number of LLC (last-level cache) misses. However, the counters are available and continue to produce values, moreover the same counters on other processors work correctly, so the user of the monitoring system may not even understand that incorrect data will be output in this case. Without the data correctness verification, it can be very difficult to detect such issues.

On Lomonosov-2 supercomputer [21], we use the DiMMon monitoring system. And we want to get information on the activity of data transfer over the MPI network using this system. To do so, we collect the amount of received and transmitted data from the Infiniband network card on each compute node. According to the Infiniband standard [4], such data is stored in units of 32 bits, i.e. 4 bytes. Thus, it was necessary to multiply the raw monitoring data in this case by 4 in order to obtain values in the conventional units of bytes/sec. However, a mistake was made at this stage, and multiplication by the wrong constant was performed. In this case, checking the data correctness, for example using the proposed tests, would immediately show that the results obtained are incorrect.

In some cases, during monitoring system setup, it is not always obvious which of the available performance monitoring counters will best reflect the desired performance characteristic. For example, when setting up the DiMMon monitoring system on Lomonosov-2 supercomputer, it was not possible to directly collect information on the number of L1 cache misses (since there was no such hardware counter available on Haswell processors used in Lomonosov-2, unlike the case of last-level cache). But the L1D:REPLACEMENT counter was available, which "counts the number of lines brought into the L1 data cache". Based on this description, there was no complete certainty that this counter fully corresponds to our needs, i.e. that it will accurately reflect the number of L1 cache misses, maybe due to the prefetcher or other peculiarities of memory subsystem. We know that this counter is used in PAPI for collecting the number of cache misses, but we would like to be sure that it is accurate in our particular case. Verification is also well suited for this purpose.

In performance analysis, it is often necessary to perform different transformations on the raw monitoring data. Let us consider one of the basic performance characteristics of HPC applications – CPI (Clockticks per Instructions) [1]. To calculate it, you need to collect two types of raw data – the number of unhalted processor cycles (clockticks) and the number of instructions retired, and then divide one by the other. And if, for example, we are interested in analyzing the state of the supercomputer for the last month, only one average IPC value needs to be calculated for each executed application. This means that for each application, the collected IPC data needs to be aggregated both by time (monitoring data is collected at a certain frequency, e.g. once a minute) and by compute nodes. Note that these transformations are usually performed outside of the monitoring system. In this case, errors can be made both when performing the transformations themselves, and when choosing the order in which they should be performed. The same is true for other collected performance characteristics as well. Data verification can be helpful in this case.

Therefore, in our opinion, it is useful to check the correctness of the monitoring data obtained for the HPC performance analysis.

4   Description of the Test Suite

The architecture of the proposed software solution is as follows. Each test is implemented as a separate C or Python module and is called independently. Separately, a wrapper was developed that allows you to set up and run all the needed tests using one command. It is possible to run tests directly or using

Slurm resource manager (via `sbatch` command). Using this wrapper, the input parameters for each test can be specified if necessary. For example, it is possible to specify the time of test execution or utilization level for some tests. Also, this wrapper allows, after the completion of all test launches, to automatically parse the output of all tests and provide a single small output file with all the results of interest.

The tests are run on one compute node (two nodes in the case of MPI tests) and check the data from that node only. There are two remarks worth noting here. Firstly, this set of tests can be run multiple times, which allow checking the correctness of data from different nodes. Secondly, the agents on different compute nodes are most often absolutely similar, and therefore checking one of them will most likely show whether the data collection is generally correct, at least within one partition.

A total of 10 tests have been implemented so far, which allow checking the following performance characteristics (collected for the entire compute node, not just one core):

- CPU user load, load average;
- I/O read and write rate;
- GPU load;
- frequency of L1 and LL (last-level) cache misses;
- amount of free RAM memory;
- data and packets transfer rate over MPI network.

It is important to note that all tests in the test suite verify the values of performance characteristics at the software level. So, they do not impose any restrictions on how exactly the performance monitoring data is collected, i.e. what monitoring system is used and data from which system sensors it obtains.

Four of the aforementioned tests were implemented using existing solutions. To assess CPU user load and load average, we use `cpuoccupy` anomaly test from HPAS suite. This anomaly performs arithmetic operations on random numbers in a loop and then sleeps for the specified percentage of the time, thereby generating the required utilization level. Since in our case we are interested in loading the entire compute node, we run this test on each of the available cores. The amount of free RAM memory is checked using the `memleak` HPAS anomaly that generates controlled memory leak. Running this anomaly for a certain time, we can estimate the difference between free memory at the start and end of this test, thereby we can check if this difference is the same as shown by the monitoring system. MPI data transfer rate is assessed using `osu_bw` benchmark from OSU Micro-benchmarks test suite. It works on two nodes, with the sender node sending out a fixed number of MPI messages to the receiver and then waiting for a reply from the receiver.

Other tests were developed within the framework of this work. Next, we will describe how several of them are implemented.

4.1   Frequency of Cache Misses

Here, we describe two tests aimed to check L1 and LL cache miss rates. These performance characteristics are usually collected by monitoring systems using performance monitoring counters (e.g. via PAPI) and are needed to assess the memory usage efficiency.

The tests work in the following way. First, the length of the cache line and the total cache size of the desired level are automatically determined using system constants, for example, _SC_LEVEL1_DCACHE_SIZE to get the size of L1 cache. Next, three arrays are created that are much larger than the cache size. Further, in the main loop, the operation A[i] = B[i]*C[i] is constantly performed, where i is corresponds to the following order of array elements: the first element is randomly selected in the range from 0 to the length of one cache line, and each subsequent element is taken randomly in the range from one to two cache line lengths from the previous position (see Fig. 1). If the program reaches the end of the array, the actions start from the beginning. Since the size of the array greatly exceeds the size of the cache of the desired level, this means that the beginning of the array has already been evicted from the cache and another miss will occur.

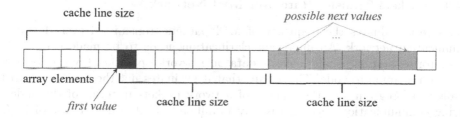

Fig. 1. The scheme of selecting next array element in cache miss tests

Accessing array elements in such a way results in a constant cache miss. Therefore, the cache miss rate in this case is defined as the total number of accesses to the array elements divided by the test execution time. These tests do not depend on manually defined constants and can be applied without modifications on different modern processors.

4.2   I/O Read and Write Speed

In this section, we describe two more tests that check the speed of reading and writing from files.

In modern supercomputers, there are often no local disks on compute nodes, so all work with files is done using a distributed file system and requires data to be transferred over the communication network. For example, Lomonosov-2 uses the Lustre file system, and a separate network is used to work with it. At the same time, we note that the proposed tests do not impose restrictions on how exactly the work with the file system is organized and will work correctly even if local disks are available.

The I/O read test is written in Python and works as follows. At the beginning of its execution, the current time is measured, then 1 GB of data is written to a temporary file. Then, this file is read in a loop, and the total amount of bytes read is stored. This continues until the program execution time specified at startup expires. When the specified time is reached, the total number of bytes read is divided by the number of seconds elapsed, so we get the average read speed, in bytes/sec. It is worth noting that the number of seconds elapsed includes the time spent writing data to a temporary file. This is necessary because monitoring systems provide information on the entire job, i.e. they include the entire duration of the test, including the preliminary stage of creating the file.

This test allows changing the execution time and the location of the temporary file. By default, the program execution time is 10 min, and the file storage folder is the current folder.

The write test is similar. In a loop, the data is written to a temporary file of 512 MB or more (calculated automatically, depending on the write speed). When the specified test execution time is reached, the total number of bytes written is divided by the loop execution time, which gives us the average write speed in bytes/sec. The launch parameters for this test are similar to the reading test.

4.3 Packets Transfer Rate over MPI Network

This test evaluates the frequency of MPI packets transmitted over the communication network. An important clarification needs to be made here. HPC monitoring system usually collects data on network operation from a network card on a compute node. This means that it evaluates not the number of MPI calls in a program, but the number of network packets in terms of the underlying communication network (usually Infiniband or Ethernet in case of HPC systems). For example, if the program includes MPI_Send call which sends 100 MB of data, this data will be split on the network level and sent using several network packets. And the monitoring system will take into account exactly this number of network packets.

Therefore, we had to write an MPI test in such a way that we could accurately estimate the number of network packets. The easiest way to achieve this is to make sure that there is exactly one network packet per MPI call. To do this, you need to make sure that: 1) the data used in one MPI call is not split between several network packets; 2) there is no delay, i.e. the data is sent right after the MPI call was made; 3) data from several MPI calls is not combined for sending within one network packet. The first condition is fulfilled if we send one byte in each MPI call. To fulfill the second and third conditions, we need to organize a synchronous data transmission and be sure that the data sent by one process was received by another process before starting the next MPI send operation.

With this in mind, a C+MPI test was written in which two MPI processes sequentially transfer one byte of data to each other, using MPI_Ssend operations. The operation scheme of this test is shown in Fig. 2.

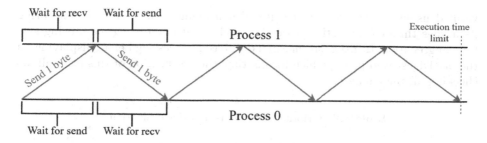

Fig. 2. The scheme of data transfer between processes

When completed, this test measures the execution time and the number of MPI send operations performed, which gives an estimate of the average number of packets transmitted per second.

5   Evaluation of the Proposed Solution

The developed test suite was firstly approbated on the laptop (its characteristics are shown in Table 1), in order to check that the tests themselves work properly. We checked everything except MPI tests, since the laptop has only one CPU. All test results were manually compared with the values collected directly from system sensors in real time. `htop` command was used to check CPU load, load average and free memory tests, `dd` for I/O tests, `nvidia-smi` for GPU load test and PAPI library for cache miss tests. Each test was executed in a usual user mode, experiments were repeated several times in order to be sure that the results are stable. For each test (i.e. for each performance characteristic), we calculated the difference between expected and observed values as |`monitoring_value*100/expected_value - 100`|, where `expected_value` is the value provided by the test, and `monitoring_value` is the value obtained directly from system sensors. This approbation showed that the average difference for each test is less than 2%, which confirmed that all tests work correctly (such accuracy is suitable in our case, see below).

After that, the proposed test suite was evaluated on two supercomputers installed in the Lomonosov Moscow State University — Lomonosov-2 and IBM Polus [2]. The characteristics of its compute nodes used for evaluation are shown in Table 1.

In case of Lomonosov-2, the evaluation was carried out the following way. Test suite was launched in the usual user mode, using the Slurm resource manager. Several runs were made to ensure the repeatability of the results. To obtain data from the monitoring system, the report generation system called JobDigest [15] was used. This system, which is used by Lomonosov-2 system administrators on an ongoing basis, allows you to automatically build reports on the performance and behavior of any executed job. JobDigest report provides a variety of data about the job operation, including all the monitoring data that can be verified using the proposed test suite. When using JobDigest, the monitoring data goes through the entire processing cycle (collection using monitoring system, aggregation, transfer, storage in a database, processing before displaying on the JobDigest web page), which means that the verification of this data allows checking all these stages.

Table 1. Description of systems used for evaluation

	Lomonosov-2 node	Polus node	Laptop
CPU	1x Intel Xeon E5-2697 v3	2x IBM POWER8	1x Intel i5-10210U
GPU	1x NVIDIA Tesla K40s	2x NVIDIA Tesla P100	1x NVidia GeForce MX350
RAM	64 GB	256 GB	8 GB
OS	CentOS 7	Red Hat 7.5	Ubuntu 20.04

To check the data, a manual comparison was made between the expected values (given by the tests after execution on the supercomputer) and the real-life values (given by the JobDigest). The achieved difference is shown in Table 2.

Table 2. Difference between expected values and monitoring data values

Performance characteristic	Difference (%)
CPU user load	0.01
Load average	0.30
Speed of I/O read	0.04
Speed of I/O write	0.14
GPU load	1.22
Frequency of L1 cache misses	0.28
Frequency of LLC misses	1.12
Amount of free RAM memory	0.39
Data transfer rate over MPI network, bytes	1.62
Data transfer rate over MPI network, packets	1.11

In all cases, the difference turned out to be less than 1.7%, which indicates the correctness of the data provided by the monitoring system. This is not surprising, since performance analysis on this system has been being performed for many years, and previously discovered issues (see Sect. 3) have been already fixed. Such difference can be considered insignificant, since the monitoring data collection process depends on many factors, such as the activity of the supercomputer system software running in parallel, jobs of other users performed simultaneously, etc. Therefore, identical runs of the same job may show slight differences in performance data due to the above factors.

In case of IBM Polus, there was no monitoring data available. Since the main purpose of the test suite is to check the correctness of the data collected for the performance analysis using the monitoring system, which is not present in this case, this evaluation was performed rather to additionally verify the correct operation of the test suite itself, as in case of laptop evaluation. All test results were again manually compared with the values provided by the operating system and system libraries in real time, using the same methods as in laptop case, plus PAPI was used for checking packet transfer rate over MPI network. We were not able to run GPU load test and test for MPI data transfer rate in bytes due to technical issues. In all tests performed, the difference in values was not higher than in the case of Lomonosov-2.

6   Conclusions and Future Plans

In this paper, we present a test suite that allows checking the correctness of the performance data collected on supercomputer nodes by the monitoring system. To do this, each test, when executed on the target system, produces expected values for a specific performance characteristic, which can then be compared with the real-life monitoring data. This solution can be easily ported to other systems and expanded with new tests (10 tests for commonly used performance characteristics are currently implemented), so we hope it can be useful for administrators of different supercomputer centers.

In the future, we plan to expand the functionality of the developed software. We are going to add new tests to verify more performance characteristics, as well as develop methods for automated verification, i.e. comparison between expected values provided by the test suite and observed values obtained using the monitoring system.

Acknowledgement. The results described in this paper were achieved at Lomonosov Moscow State University with the financial support of the Russian Science Foundation, agreement No. 21-71-30003. The research is carried out using the equipment of shared research facilities of HPC computing resources at Lomonosov Moscow State University.

References

1. CPI description. https://www.intel.com/content/www/us/en/develop/documentation/vtune-help/top/reference/cpu-metrics-reference.html#cpu-metrics-reference_CLOCKTICKS-PER-INSTRUCTIONS-RETIRED-CPI
2. Description of IBM Polus supercomputer (in Russian). https://hpc.cmc.msu.ru/polus
3. OSU Micro-benchmarks. https://mvapich.cse.ohio-state.edu/benchmarks/
4. InfiniBand Architecture specification, volume 1, release 1.3 (2015)
5. Intel Xeon Processor E5 v2 Product Family. Specification Update. September 2015. Tech. rep. (2015). https://www.intel.com/design/literature.htm
6. Intel Xeon Processor E3–1200 v3 Product Family. Specification Update. October 2016, Revision 016. Tech. rep. (2016). https://www.intel.com/design/literature.htm

7. Ates, E., et al.: HPAS: An HPC performance anomaly suite for reproducing performance variations. In: Proceedings of the 48th International Conference on Parallel Processing, pp. 1–10 (2019)
8. Barth, W.: Nagios: System and network monitoring. No Starch Press (2008)
9. Bautista, E., Romanus, M., Davis, T., Whitney, C., Kubaska, T.: Collecting, monitoring, and analyzing facility and systems data at the National Energy Research Scientific Computing Center. In: ACM International Conference Proceeding Series. Association for Computing Machinery (aug 2019). https://doi.org/10.1145/3339186.3339213
10. Brandt, J.M., et al.: Ovis-2: A robust distributed architecture for scalable ras. In: 2008 IEEE International Symposium on Parallel and Distributed Processing, pp. 1–8. IEEE (2008)
11. Evans, T., et al.: Comprehensive resource use monitoring for hpc systems with tacc stats. In: 2014 First International Workshop on HPC User Support Tools, pp. 13–21. IEEE (2014)
12. Gruber, T., Eitzinger, J., Hager, G., Wellein, G.: LIKWID 5: Lightweight Performance Tools (2019)
13. Kluge, M., Hackenberg, D., Nagel, W.E.: Collecting distributed performance data with dataheap: generating and exploiting a holistic system view. Proc. Comput. Sci. 9, 1969–1978 (2012)
14. Massie, M.L., Chun, B.N., Culler, D.E.: The ganglia distributed monitoring system: design, implementation, and experience. Parallel Comput. 30(7), 817–840 (2004)
15. Nikitenko, D., et al.: JobDigest - Detailed System Monitoring-Based Supercomputer Application Behavior Analysis. In: Supercomputing. Third Russian Supercomputing Days, RuSCDays 2017, Moscow, Russia, September 25–26, 2017, Revised Selected Papers. pp. 516–529. Springer, Cham (sep 2017). https://doi.org/10.1007/978-3-319-71255-0_42
16. Olups, R.: Zabbix Network Monitoring. Packt Publishing Ltd (2016)
17. Palmer, J.T., et al.: Others: open XDMoD: atool for the comprehensive management of high-performance computing resources. Comput. Sci. Eng. 17(4), 52–62 (2015). https://doi.org/10.1109/MCSE.2015.68
18. Shvets, P.A., Voevodin, V.V.: "Endless" workload analysis of large-scale supercomputers. Lobachevskii J. Math. 42(1), 184–194 (2021)
19. Sottile, M.J., Minnich, R.G.: Supermon: A high-speed cluster monitoring system. In: Proceedings of IEEE International Conference on Cluster Computing, pp. 39–46. IEEE (2002)
20. Stefanov, K., Voevodin, V., Zhumatiy, S., Voevodin, V.: Dynamically reconfigurable distributed modular monitoring system for supercomputers (DiMMon). Proc. Comput. Sci. 66, 625–634 (2015). https://doi.org/10.1016/j.procs.2015.11.071
21. Voevodin, V., et al.: Supercomputer Lomonosov-2: Large scale, deep monitoring and fine analytics for the user community. Supercomput. Front. Innov. 6(2) (2019). https://doi.org/10.14529/js190201

Author Index

© The Editor(s) (if applicable) and The Author(s), under exclusive license
to Springer Nature Switzerland AG 2023
V. Malyshkin (Ed.): PaCT 2023, LNCS 14098, p. 209, 2023.
https://doi.org/10.1007/978-3-031-41673-6

Printed in the United States
by Baker & Taylor Publisher Services